Big Brains and the Human Superorganism

Big Brains and the Human Superorganism

Why Special Brains Appear in Hominids and Other Social Animals

Niccolo Leo Caldararo

ROWMAN & LITTLEFIELD
Lanham • Boulder • New York • London

Published by Rowman & Littlefield
A wholly owned subsidiary of The Rowman & Littlefield Publishing Group, Inc.
4501 Forbes Boulevard, Suite 200, Lanham, Maryland 20706
www.rowman.com

Unit A, Whitacre Mews, 26-34 Stannary Street, London SE11 4AB, United Kingdom

British Library Cataloguing in Publication Information Available

Library of Congress Cataloging-in-Publication Data

Names: Caldararo, Niccolo Leo, author.
Title: Big brains and the human superorganism : why special brains appear in hominids and other social animals / Niccolo Leo Caldararo.
Description: Lanham : Lexington Books, 2017. | Includes bibliographical references and index.
Identifiers: LCCN 2017033832 (print) | LCCN 2017023021 (ebook) | ISBN 9781498540889 (Electronic) | ISBN 9781498540872 (cloth : alk. paper)
Subjects: LCSH: Brain—Evolution. | Primates—Evolution. | Cognition.
Classification: LCC QP376 (print) | LCC QP376.C27 2017 (ebook) | DDC 612.8/2—dc23
LC record available at https://lccn.loc.gov/2017033832

™ The paper used in this publication meets the minimum requirements of American National Standard for Information Sciences—Permanence of Paper for Printed Library Materials, ANSI/NISO Z39.48-1992.

Printed in the United States of America

To my wife, Ann Sheldon, who helped me traverse the complexity of the brain.

Contents

List of Illustrations

FIGURES

TABLES

Acknowledgments

I would like to thank Ralph Jones who has guided this book through every stage of the editing process with good humor and elegant expertise. When all seemed insurmountable, he steadily and professionally carved away at the brambles of detail to give a clear path. Also I must thank the staff of Lexington Books for their help and patience in the development of the text, and several anonymous reviewers whose comments and criticisms greatly improved the text.

Part I

BRAINS AND PERFORMANCE

Chapter One

Cranimania and Human Behavior

Yes, education is a treasure and culture never dies.

(Petronius, 1913)

The search for an explanation of the human brain's evolution has had a varied history. Holloway (1970) has argued that our primary problem in understanding brain and behavior evolution is not the growth or expansion of cranial capacity. He is right, and the focus on cranial capacity is indicative of a basic conundrum: Why do we have big brains, and why did they evolve? In general, the big brain has been seen as a result of skill, tool making, social interaction, cooperation, and competition. Krantz (1968) argued that the need to pursue an animal over long periods of time (he called it "persistence hunting") required of early hominids the ability to maintain the image of the kill. Of course, other animals, such as wolves and hyenas, do this (Mech, 1970; Smith et al., 2012), and the pursuit would seem special in hominids primarily in terms of energy expended. Unfortunately, this theory requires us to project contemporary hunter-gatherers into the past as early hominids. Yet, even proponents of the theory, such as Liebenberg (2006), can report only a few cases of this method, and, when exceptional cases were followed, Liebenberg reports a success rate of only 50 percent. He proposed that it was associated with hunting at the hottest part of the day, and thus, his idea could be related to Falk's (1990) "radiator theory," discussed in more detail later. There are a number of physiological complications in comparing heat loss and energy in mammals that complicate this scenario (Carrier, 1984). Left unclear, especially in light of other carnivores chasing prey across various distances, is how "far" qualifies for selective pressure, how often and under what conditions. Determining the beginning of a chase sequence in wild predators is difficult, even when using electronic tags, yet we have data on different pursuit strategies for some, as in wolves and some felids. The success rate for wolves who pursue prey over long distances is not significantly better than that reported by Liebenberg (2006) for humans (Husseman et al., 2003).

Neel (1970) argued that the main feature separating hominids from other primates was population control, which could be considered an element of social complexity. Some economists have argued that the use of coal created modern economy, as if humans had not used coal before (Gowdy and Krall, 2016). Yet, the Romans and the Chinese used coal (Freese, 2004; Smith, 1997) and there is evidence the Aztecs also did (Freese, 2004). The industrial revolution is more a product of a combination of colonialism, slavery, and European runaway population expansion after the Black Death combined with discoveries in public health, especially water-borne disease (Caldararo, 2012b). Isler and van Schaik (2012) introduced a similar argument recently. But Hrdy (1974) found langur monkeys to engage in similar behaviors (mainly infanticide), though Jay Dolhinow (1963, 1965) had contradictory findings, and the matter remains controversial (Opie et al., 2013) but may be associated with the development of social monogamy. The main problem with these theories is that they fail to link expensive brain tissue with reproductive success and can seldom be linked to evidence in the fossil record. Hrdy (1999) has produced a modification of this approach, but the link to a necessity for large brains is still lacking. Another related approach is to use the metabolic turnover, the basal metabolic rate, and brain size of an animal. There is a pronounced correlation in primates, though the fact that chimpanzees and humans are similar in predicted value (Isler and van Schaik, 2012) undermines the explanatory significance of the correlation, and references to relative body fat (Wells, 2010) do not add significant explanatory power to the thesis. One might argue that the nature of evidence of complex behavior is not preserved in the fossil record until 10,000 BP—that is not an answer, yet it presents a question: Why? It seems that part of the problem lies in disagreements over what complex behavior is, and how it can be discovered in the fossil record.

The human nervous system contains some 100 billion neurons (Azevedo et al., 2009), that of the short-tailed shrew about 52 million (Hofman and Falk, 2012), the common tree shrew 260 million (Herculano-Houzel et al., 2007), the brown rat about 200 million (Herculano-Houzel and Lent, 2005), the tufted capuchin 3.6 billion, the Rhesus macaque 6.4 billion, octopus over 500 million (Hochner et al., 2006), bees around 950,000 (Menzel and Giurfa, 2001), and various ant species around 250,000—and one assumes that more neurons mean more processing, which provides an advantage for survival.

The cerebral cortex of humans contains some 20 billion neurons, whales from 10 to 40 billion, African elephants 11 billion, chimpanzees about 6.2 billion, capuchins around 600 million: variation in numbers is substantial in different species and individuals (Herculano-Houzel et al., 2015). One can be overwhelmed by the massive complexity of mammal and bird brains, but given the expense of maintaining the living needs of an organism's network of cells, the question of the need or use for complexity becomes paramount.

The amount of fat (lipids of various kinds) in the human brain is similar to that of other mammals, yet water makes up the greater portion. One sometimes comes across references to the brain containing 60 percent fat, but this is a mistaken reading of O'Brien and Sampson's (1965) report on an analysis of residues after the evaporation of water. Much of this fat is in myelin, which acts to increase the efficiency of transmission of impulses from one neuron to the next (Hartline, 2008). Myelination of neurons differs in species across mammalian orders, and there is some disagreement concerning the amount of myelination in certain big-brained species such as dolphins (Marino, 1996; Marino et al., 2007). Though Manger (2006) argued that large cetacean brains evolved in response to a decrease in water temperature in the Eocene, Marino et al. (2007) demonstrate that the evidence, which points to selection for complex cognition, does not support this claim. So, one might say that we have fat in our brains to increase the efficiency of communication between our brains and our bodies' cells, with the goal of interacting effectively with the environment. But then, why are there differences in both the number of neurons in the brain and the brain size? How does increased brain size relate to success and survival, and in particular, the survival of humans? Striedter (2005) quotes van Dongen (1998) implying that this is a near-impossible task: "too many factors seem to be associated with brain size in primates." This question, and the difficulties in providing an answer, will be partly discussed in part I of this book in direct relation to the evolution of hominids, and then in a more specific manner regarding human evolution in part II.

Other vertebrates with large brain-to-body ratios, such as dolphins, also engage in significant social interaction, but this has not resulted in substantial reproductive success: Dolphins are minor actors in almost all aquatic environments and certainly have not had the dominance in their ecological settings that humans have. Nevertheless, some researchers have linked social complexity with intelligence and big brains in both humans and dolphins, without manifestations of material developments or environmental dominance (Bearzi and Stanford, 2010). Human behavioral complexity is also relatively new, and human ecological dominance has only arisen since the development of domestication and sedentary life. Parallels with the social insects are obvious in the building of structures, war, slave making, domestication of plants and animals, social hierarchy, and learning. The social brain theory is an obvious answer. Original research with nonhuman social mammals such as beavers (Campbell et al., 2005) and prairie dogs, in terms of brain size, was used as a means to test this theory and is included in this book. What differences are there between the brains of rodents that are not social and those of highly social rodents? Prairie dog crania and beaver crania were measured and compared to published data and are presented in this book.

Brains from highly social rodent species are larger than those from non-social rodents, and the results of this study seem to support the social brain theory. But reductions in human brain size following the achievement of components of eusociality seem to suggest a domestication effect. Recent neurological findings of analogies in specific neuronal conditions in ants and humans provide new paradigms for understanding the evolution of the human brain and eusociality. We address the problem of defining complex society later (see also figure 8.1, where I make a selected listing of elements of eusociality and associate them with general groups of animals).

These findings are placed into the context of assumptions that have been put forward in a number of theories explaining the reason for the evolution of big brains. Often, this has manifested as a fixation with the human cranium in size and/or shape, what Clemence Royer called the "skeletonmania" of anthropology, or from Antenor Firmin's (1885) perspective, a "craniomania." We dissect what brain size and cranial shape have to do with what it means to be human. This work follows on earlier research conducted on the idea of unique human physiology and behavior (Caldararo, 2009a, 2012b) and a study placing human ecology in a uniformitarian context (Caldararo, 2013), that is, all of what humans do see as part of a continuum of animal behavior.

The second part of this book examines the evolution of human consciousness as well as the idea of human cognition and mind in the context of physical means of function, that is, electrons, bonding, complex molecules, and synapses. These physical means have produced a feature of life we call humankind, with a uniform response to the environment. By this I mean "uniform" in terms of "seeing" by mediating part of the spectrum of light, learning to respond to physical states like greater or lesser amounts of heat, and possessing a biological framework, "mind," that presents each individual with a map of responses to these stimuli. We call this map "culture." Then, I am going to discuss whether "mind" can exist without "culture" or "language." Can one be human without language? Or is mind a creation of environmental stimuli in reference to the organism, retained and reused as brief encounters in the manner that Greenfield (2016) argues for "neuronal assemblies" which evolve into a collective and continuous experience of consciousness, something Reber (2016) believes is much more common in an evolutionary sense? Since this implies learning at the cellular level, and interaction of neurons in neural nets is ancient in the history of multicellularity (Arendt et al., 2016), one would imagine so would be "mind" or "consciousness."

This second section also examines how the ways humans have lived, including the effects of disease, have formed the genetic background of modern humans and created a specific kind of adaptive creature from the context of hominid evolution over the past 10,000 years, and how such factors likely have had a role in the development of sociality in other animals (Danforth, 2002).

We will discuss this transition in some detail in later chapters. It began when sedentary life was invented, along with domestication of plants and animals, from the earlier collecting platform of behavior that evolved from the Primate heritage beginning with Sahelanthropus over seven million years ago. The second part also addresses the problem of mind and how human attempts to understand mind are tied to ideas of cosmology, humans as the center of creation, gods' meanings, and the "reason" for the perceptions we have such as the universe of stars. What can we learn about reality from this brain that can relate the life we find on Earth to that on the stars, and in fact is what we call life, so interrelated that we cannot separate ourselves from it? Will we, like Carl Sagan, find our neurons like stars, within a limited tangle of electrical impulses traveling on axons amid nucleoproteins, all like gravity and dark matter interacting in complex patterns whose interactions make "life" or "space?" As Dr. Alan Mann observed in a seminar at UC Berkeley in 1970, we have a Stone Age mentality controlling a space age technology, and there will be no "safety net" if the universe is, as Albert Camus once said, indifferent to our existence.

Central to our understanding of what the brain does for us is the effort to examine learning and its context, as well as the costs of learning and technology. Chapter 11 will explore some of these problems, such as the fact that learning environments in the past were generationally based, with one group of adults teaching children. This has been true for almost 99 percent of human history. Today, we are increasingly learning from technology—television, computers, smartphones (which, of course, are computers)—and this learning takes place in different contexts: alone, in groups (class, workshops, etc.), and in virtual assemblies. Some scientists and health care professionals think that the way we are living is creating learning disabilities like ADD/ADHD, dyslexia, and autism spectrum disorders (Donovan and Zucker, 2016; Perou et al., 2013).

It is unlikely that the traditional school setting has changed dramatically from early complex societies, for example, ancient Egypt or Han China (Dawson, 1925; Henson, 1980; Reischauer and Fairbank, 1958), but most from nonindustrial society (Kenyatta, 1938). Descriptions of the educational setting in ancient Rome (Bonner, 1978) seem similar to those in the nineteenth century of rapidly industrializing England. What is not surprising is that the educations of the children of rich and poor are still remarkably different in the twenty-first century, as they were in the nineteenth century, or the feudal past (Sojberg, 1965). A study by the Pew Research Center (2015) found that class differences in child rearing are growing, a symptom of widening inequality.

Just as caste differences in India are based on restrictions in education, nutrition, and opportunity, the creation of caste in the West seems to be underway.

If caste and division of labor are considered the driving forces in eusociality, one wonders why insect societies show such dramatic avenues of plasticity, especially in the lives of individuals who transition into different types of work in a lifetime (Roisin and Korb, 2011). Development into a caste in humans is also not specifically biologically determined, but conditional, as in termites. In humans, if caste and the division of labor were so effective and efficient, one wonders why the caste system in the Indian subcontinent has not produced a more efficient and dynamic society, why development has been so limited, and why it still is uneven (Jeffrelot, 2005). It would appear that the division of labor is dependent on another factor, one that facilitates plasticity and produces a varied degree of adaptability. Specialization and derived characters often lead to extinction, while generalization allows more opportunity—the caste system in India being an example of hyper-specialization (Dumont, 1966). As Berreman (1978) shows, the density of population with surplus and specialization can lead to caste and class. But in human society, we have found that this process of specialization can become a liability with the growth of bureaucracies that parallel the growth of retired and unproductive citizens. In the late Roman Empire, Lactantus complains to Emperor Diocletian of his expansion of the army and civil service (Cipolla, 1970). Cipolla (1970) argues that "military expenditure powerfully contributes to the growth of total public consumption." Rostovtzeff (1926) shows how, in the Roman example, this leads to civil war, military dictatorship, and the collapse of law and order.

We might imagine that the process of self-domestication has set us on the road to the kind of animal society, characterized by increased complexity and specialization, that Wilson (1975) has described for some insects. We must keep in mind that humans are the only animals that have self-domesticated, become food producers, and eliminated all their predators. Though cooperative defense against predators was not found to correlate with larger brains (Smith et al., 2012), the evolution of sociality may still arise as a defense against predation, and we find that parental care is associated with an attack on young in both the prey and the predator (Magalhaes et al., 2005). The finding that evidence appears in the Early Cretaceous of advanced sociality in termites with their most serious enemy, ants (Engel et al., 2016), supports this idea. We might note that insect eusociality was a successful adaptation to many challenges but failed to eradicate the species' predators, while humans appear to have achieved this at the end of the Mesolithic and the dawn of the Neolithic. Pellens et al. (2007) note that predation can cause a reversal in eusociality in cockroaches, and Hart et al. (2005) argue that predation has been a driver of eusociality in hominins.

Every now and then, an alligator or tiger kills a human, usually a child or person unaware, but we are not threatened in general by such animals. Our teaching has changed from an individual context to a mass context, parallel to the mass media that have gone from scrolls and parchment, clay tablets, or newsprint to machines like computers. Thus, we might consider that evolution to the kind of "superorganism" that Hölldobler and Wilson (2008) have described is on its way.

In economics, when the "fundamentals" concerning production, consumer behavior, and management indicate that a stock should go up, it often does not, and when all data provide a picture of disaster, and the stock should fall, the market surprises. All the desires of consumers, managers, investors, financial brokers, and analysts combine to produce what Kroeber (1917) defined as the sum total of human thought and action, the "superorganic" (different from that first proposed by Herbert Spencer, 1876). Man has created a god that circles the globe treating man's billions of needs, a god made not in man's physical image but in the visage of his primitive psychology and peculiar neural calculus, formulated, as Leslie White (1947) proposed, within the human mind's experience of history. This human "superorganic" has marked similarities with that of Hölldobler and Wilson's insect superorganic, as we shall see in more detail later in this book. The failures of financial analysts and economists to predict the market (Coggin, 2006) and the comments of Graham Copley, the HSBC global head of equity research, that analysts are "worthless" (*Financial Times*, June 19, 2006) are no different than the failure of shamans and priests to intercede with a deity to produce the desires of their worshipers (Eliade, 1964). This includes the exposés of insider trading and conflicts of interest that analysts were involved in, since shamans often were acting in their self-interest as well, and brokers need to buy and sell for clients to make a living.

In examining this process of evolution of the superorganism, we have to realize that, as our sheer numbers have increased and the density of our living arrangements become greater, the needs for providing for these masses must change the way humans interact. Berreman (1978) has shown this in India, where increased density of population results in increased rigidity of caste. Discipline and order are achieved more efficiently through education and communication to increase the conditioning schedule effectiveness. Shorter and faster signals, delivered with more abstract references and sent out on microwaves like texting, can influence millions—and perhaps billions—in ways human society has never achieved before (Boyd, 1915; Gardner and Davis, 2013; Tucker, 2014). Some ant species engage in war, as do some bees who raid others' colonies and inflict surrender or submission (Nieh, 2010). Organizing others to build, fight, or surrender requires effective forms of communication. Surrender and conversion are complex behaviors and will be discussed later.

Whether this is "good" or "bad," the evolution of the ability of a society to deliver food and other resources to support its population must keep pace with needs or the society will disappear. I have previously discussed both these issues in the culture history of human economic society (Caldararo, 2013); here, we will address them more specifically in regard to the present future of globalism and capitalism, given that it appears capitalism has won the great contest with socialism, which raged across the last two centuries. My caveat to this is in my speculations on Japanese communitarianism (Caldararo, 2003b, 2014) and Melanesian contact with Westerners and the potential for positive change these examples suggest.

Creativity and innovation are central to human feelings of uniqueness (Cipolla, 1978; Williams, 1987). We will address this in a number of contexts, but the problem of definitions will constantly intrude—for example, not only the idea of creativity but also other more basic concepts like plasticity and "unconventional" behavior. Striedter (2005) suggests that impulse control is at the base of such behavior and is a consequence of developments in the prefrontal cortex, and since such behavior is seen in some birds' (e.g., crows) brain segments, the DVR is an analogue. We will examine these tendencies to juxtapose brain evolution with behavioral categories. Obviously, this is important as it touches on one of the foremost ideas of human behavioral plasticity, and is a central component of the current economic ideology that humans are innovators. The idea of novelty is not lost on other animals, and many cultures have imputed it to certain animal characteristics, as in "curiosity killed the cat." Cultures vary in their sensitivity to the effects of the new or the ways in which new ideas penetrate and become established, often a process we call appropriation or acculturation. Japan's experience with the West (Caldararo, 2003b) and Melanesian contact with Westerners are good examples of this process (Mead, 1964).

One section of this book, chapter 11 on ADD/ADHD, has been published in part previously, but the present chapter is more than 80 percent new material. It contains parts of an article previously published in *American Anthropologist*, but the material is relevant here to our discussion as it addresses central issues in human evolution regarding learning and the sick role. I have written extensively on human evolution, speciation, and DNA analysis, especially studies of ancient human DNA. These ideas are central to what it means to be human, how we regard difference and especially normality. What happens when someone is functional and then not, changed by disease or trauma? What happens to their humanity? What "difference" means, and how it developed, is key to understanding human diversity and the nature of humanity. Chapter 12 looks at how we have lived over the past seven million years, and it focuses on the ways that changes in how we live and use space have affected the way we live, and also the idea of homelessness. The last chapter, chapter 13, looks at the future of cognition and brain to find, from what has been presented, an image of the way our thinking can guide human survival and that of the planet.

Chapter Two

Brains

What Are They Good For?

Regarding the development of consciousness, J.B.S. Haldane observes: "The strangest thing about the origin of consciousness from unconsciousness is not that it has happened once in the remote past, but that it happens in the life of every one of us" (1990). Perhaps the greatest problem with addressing the question of the evolution and function of the human brain is that it is ours. The question of how human consciousness differs from animal reactivity has been a subject of great speculation and debate. Darwin knew that slow steps with small differences separate forms of life in a hierarchy of developed potential and that this showed but a small inequality between the higher apes and man (Darwin, 1871). While more complex and differentiating views have appeared (Skinner, 1953), alternative dissenting voices have continued to be heard (Crail, 1981; Osborn, 1927; Summer, 2011). What is human consciousness and how it came to be are central questions of an epistemological nature. The fact that all humans possess the capacity for language, even those born with severe genetic defects like Down's syndrome or Savant syndrome (Buckey, 1993; Martin et al., 2009; Sternberg, 1985; Treffert, 1989), leads us to the belief that evolution has resulted in a basic apparatus that is fundamental to self-awareness and attendant on the survival of a fetus to childhood. Normal functioning within the mode of human social activity is a more difficult quantity, as is the quantification of achievement (Bouchard and McGue, 1981; Kamin, 1995a, 1995b; Mackintosh, 1998).

Yet, this thing called consciousness is often noted by its absence; take, for example, the case of Helen, a macaque that had her entire visual cortex removed and yet apparently could perceive movement (Humphrey, 1974). She lacked the "consciousness" of sight but "saw," thus vision without conscious apprehension (in a Kantian sense of intuition versus perception [Kant, 1783]). This is an example of "blindsight." Theories abound about the location of consciousness or combinations of perceptions "making" or creating it. These have been tested experimentally and have produced possible examples of repression of unconscious information by conscious activity (Jolij et al., 2005). Humphrey (2006) separates sensation from perception and believes that they act as individual systems, yet how there is an integration of sensation and perception to create mind is left unexplained in Humphrey's work.

The sense of self obviously must be associated with the development of language, yet how do we study the evolution of language except by studying language acquisition in children or pathological cases? Means to evaluate and describe an internal component of the mind ("internal language" or "I-language") had been lacking until recently, yet many researchers have rejected the idea of other animals having "language" (Hauser et al., 2002). Is either of these relevant? In other words, since language is likely to have evolved thousands, if not millions, of years ago, how can we conceive of an origin of language where the brain which it evolved from is no longer in existence? Is it just a matter of genes or is it one of structures and the evolution of regions of the brain, which seems unclear today (Striedter, 2005)? Studdert-Kennedy and Goldstein (2003) trace the elements of language, both physiological and neurological, with the nature of gesture and vocal sounds in mammalian evolution and attempt to place human language within a trend of selective pressure for communication and an extension of an ancient process of all natural systems.

Crick (1995), building on ideas like "blindsight," suggested a biological framework for consciousness as a by-product of the chemical interactions of neurons, molecules, and atoms. Edelman (1978) drew a complex scheme for the evolution of conscious interaction that Fernando et al. (2008) and, later, Fernando (2013) elaborated upon, producing a more comprehensive model of consciousness.

Some biologists approach the origin of mind from the evolution of perception, based on alternative responses (presumably innate) to similar stimuli (Bonner, 1980; Griffin, 1992; Reber, 1992, 2016). Here, the complexity of response to the environment develops into systems of increasing plasticity. Some biologists, like Bonner (1980) and Alexander (1989), start this from the origins of vision with simple multiple types of response integrated (as in *The Orientation of Animals*, Fraenkel and Gunn, 1961). For Alexander (1989) the process requires a changeable memory where the visual observation of the alternatives by the organism produces a rudimentary self-awareness required for the comparison of alternatives to results. Cognitive psychologist Ulrich Neisser (1976) proposed an "anticipatory schema" to this selection process, making the response more effective and efficient, arising from the control of the activity of looking, and again forms of taxis. This is often opposed to reactive schema-based architectures of cognition, defined as quick responses to dynamic environments.

Anticipatory schema-based architectures include anticipatory mechanisms that exploit expectations of anticipated next stimuli in the environment (Seel, 2012). However, his idea that memory is a reconstructed process and not a frozen image provides a basis for understanding the development of response selection by an organism. One might argue that it is a different process to separate the image from the meaning and vocalization in the origins of language (Saussure, 1916). Graziano (2013) has elaborated on this idea with his concept of attention schema theory, where selective enhancement of images allows for concentration and focus of an animal's perceptions onto selective areas of the visual field. Unfortunately, he assigns the origins of this ability to the rise of vertebrates, which would either eliminate the demonstrated cephalopod abilities or require a case of parallel evolution. They can also use their tentacles as arms and manipulative hands (Sumbre et al., 2005), and like vertebrates, legs in walking bipedally in water (Huffard et al., 2005). The ability appears well developed in owls (Mysore and Knudsen, 2014) whose Wulst shows striking similarities to the mammalian visual cortex (Striedter, 2005). The owl Wulst is enormous compared to other birds, and its hypertrophy is similar in that regard to the human brain and primates.

To Alexander (1989), multiple-choice responses of simple animals evolve into more complex conditioning processes that are organized and developed by rehearsal in play, dreams, and "self-deception." What Griffin (1976) included in the class of "covert nonverbal communication" transferred into images projected through time, where the individual animal can "see" itself in a variety of situations. These nonreal conceptions of self and time become the nature of self-awareness. To Laing et al. (1966), the inability to separate these scenarios from reality is a basis for what is called madness. He calls human interaction "refractions" of interpretation and meaning where dyads of interpersonal assessment of intentionality are constantly going on, often in failure, yet well enough, mostly, to provide functionality for society.

Alexander's (1989) use of the term "play" is derived from that of Fagen (1981) and Piaget (1945) where play is practice, often executed as "observing in the mind" through the construction of surrogate scenarios. Huizinga (1938) had earlier devoted a monograph to the analysis of the role of play in society, emphasizing also the fact that the idea of "play" represents action. Combatants "play out" a battle, for example. Actors and priests recreate roles, some secular, others religious. Alexander (1989) adds fantasy to this process of manipulation of reality and the evolution of self, scenario building to problem-solving. One has to be careful of defining behaviors across species, as ideas of "play" can also be responses to boredom and thus reflect forms of display behavior, especially in captive animals, or self-directed behavior, sometimes described as measures of anxiety (Mason and Perry, 2000).

In another correlate to Alexander's cluster of traits for the evolution of mind, Anderson et al. (2010) claim that octopods also sleep and that there is evidence of tool manipulation and play (Tzar and Scigliano, 2003). Also, they seem to build hierarchies and engage in deception. An answer to why such short-lived animals would evolve complex cognitive abilities appears to support Steklis' (1999) suggestion that more emphasis needs to be placed on the ecological theory, which some octopod specialists find appealing (Tzar and Scigliano, 2003), though it undermines the idea that big brains evolved as a correlate of expanded life spans. Yet, studies with the cephalopods have produced indications of independent evolution of convergent means of the evolution of cognition, as Shomrat has noted (and will be of importance later when we discuss insect eusocial evolution): "The pioneering work of J.Z. Young described the morphological convergence of the cephalopod vertical lobe with the mammalian hippocampus, cerebellum and the insect mushroom body" (Shomrat et al., 2015).

One could go further: on another level of classification, squids have been found to use shells as "homes" or protective tools (Finn et al., 2009). Added to this is another group of invertebrates, social spiders (Yip et al., 2008), whose cooperative hunting seems dependent on the benefits of large prey that their group behavior provides a scaling advantage. In general, many social spider species show a predominance of female colony members (Aviles and Maddison, 1991), with significant numbers of noncolony solitary females foraging alone for presumably new colony formation in some species (Venticinque et al., 1993). Some have equal numbers of males and females (Aviles and Maddison, 1991). Colonies in some species may be divided into clusters, but genetic differentiation seems low. Some species show delayed dispersal of juveniles from the mother with certain benefits to both (Jones and Parker, 2002), an obvious parallel to socialization benefits in other animal species. Thus cooperative hunting, sedentism, cooperative building and maintenance, protection and rearing of young, and group defense of collected prey (Cangialosi, 1990) all play a role.

A more diverse example is that of ambrosia beetles (Biedermann and Taborsky, 2011) where we find a variety of eusocial developments: each female is capable of breeding, there is cooperative brood care, young are helpless, fungiculture, social construction of living areas, defense, waste disposal, division of labor, and grooming. Yet, no castes developed, and individual members perform a variety of tasks in their life history. Several different evolutionary developments appeared in this group with different mycangia (Hulcr and Stelinsky, 2016). In addition, crop theft among some species has evolved with significant deceptive strategies (Hulcr and Cognato, 2010). See figure 8.1 showing the characters associated with eusociality among different animals.

Given that mammals survived by invading nocturnal niches sometime in the Triassic and then readapted to diurnal niches at the end of the Cretaceous, their visual center must have undergone considerable modification (Striedter, 2005). As nocturnal animals' vision would have been important, so, too, would smell and auditory cues, more so than for diurnal animals, though specializations did occur as in the huge eye of the Tarsier that is larger than its brain (Conroy, 1990). If we look at the contemporary "primitive mammals" like the marsupials, we find they lack the robust connections of the hemispheres of the brain found in placental mammals and integration of various regions of the brain are not as complex (Striedter, 2005). This may reflect the less integrated state of early mammalian processing at the beginning of the Cretaceous that has become so much more so given the significant "invasion" of regions of the brain in the placental mammals as a result of the expansion of connectivity demanded by a more diverse visual environment. Hall et al. (2012) report that most diurnal and cathemeral mammals have eye shapes that are similar to those of nocturnal birds and lizards. The only mammalian clade that diverges from this pattern is anthropoids, whose eye shapes are similar to diurnal birds and lizards. Of interest here is also the problem of focus. How do animals produce a cognitive image of reality? To Sacks (1970), pathological conditions in humans can provide an avenue of clues, for example, in agnosia where an injury to the occipitotemporal border affecting the ventral stream interrupts transmission of information of either a visual or an auditory nature. Affected individuals cannot integrate stimuli; they can "see" parts, but not faces, for example.

The fact that even nanocephalic dwarves develop language (Baxter et al., 2009; Lenneberg, 1967) puts into question many theories regarding the evolution of the large human brain and language capacity. There is some debate over the assignment of cases of dwarfism, as Leviton et al. (2002) found that many specific identifications as microcephaly could be the result of periconceptual exposure, and so, serious methodological problems are associated with some of the literature on microcephaly and other dwarfism cases. Several phenotypes have been recognized in acquired or progressive microcephaly, and overlapping symptoms are present (Baxter et al., 2009). Microcephaly has also recently been associated with Zika virus infection (PAHO and WHO, 2015).

Bipedalism has been associated with brain size in the radiator hypothesis, so that the increased vascularization of the brain plays a role in cooling the body (Falk, 1990). The maternal energy hypothesis associates the mother's basal metabolic rate with its neonatal brain mass and the expensive tissue hypothesis argues that reduction in other organs allowed for greater brain growth given its costs of maintenance. But, in both these cases, Falk (2007b) argues that studies of bats have challenged the findings (especially Ratcliffe et al. [2006], concerning increased behavioral flexibility in predatory bats) while Dunbar (1998) is also negative and refers to work by Pagel and Harvey (1988). Pagel and Harvey (1990) find instead that generally (some mammals are exceptions) large brains associated with maternal metabolic rate, it is correlated with long gestation periods and small litter size. In either case, even without the work of Jones and MacLarnon (2004) with bats, we would still have Tobias' (1971) complaint that the development of the large brain is inexplicable and, we might add, even illogical.

It is likely that the idea of "big = complex" cognition will have to be reexamined, given recent research with octopods (and includes many of the Cephalopoda), where the brain is not concentrated into one main area but instead distributed in the arms and behind the eyes (Courage, 2013). But octopods also challenge the social brain theory, as they are not highly social, with rare exceptions like the Pacific striped octopus (Courage, 2014). Though their brains are larger than other invertebrates, given the body weight comparisons, much of the brain is wrapped around their esophagus and has folded lobes and memory centers (Tzar and Scipliano, 2003). Another problem for the social brain theory that octopods display is a short life span where social brain theories argue that investment in expensive brain tissue requires a long period of learning necessary to interpret intentionality in other group members and engage in efficient social behavior (Anderson et al., 2010). The exception, again, is the Pacific striped octopus that forms groups, modifies its den, couples, and shares food (Caldwell et al., 2015). Tool use has also been reported in octopods (Finn et al., 2009). Mass action may describe brain function in octopods as stimulation of different brain areas failed to produce action. Results of study of the octopod brain found, "Discrete and complex components have no central topographical organization but are distributed over wide regions" (Zullo et al., 2009). Yet, the arms possess apparently independent movement and responses to stimuli from a central processing. A similar distribution of brain segments in legs appears in spiders, yet also short-lived and highly social spiders of the genus *Stegodyphus* support arguments for the social-niche-specialization hypothesis (Laskowski and Pruitt, 2014).

Since the appearance of the Piltdown fossils, a hoax that supported the idea that big brains were necessary for a human-like hominid, the idea of the link between ever-enlarging brains and human consciousness has been ingrained in the study of the evolution of human consciousness. Ideas of the association of language with large brains have been produced, and the size of the brain has overshadowed the psychological mechanisms responsible for speech, for example, where Deacon (1997) rejects the generative grammar concept of Chomsky (1972). One aspect of Chomsky's work that is ignored is the fact that language can separate as well as communicate, that it creates domains of cognitive equality among common groups users that are blocked to those of different groups. The considerable contribution of behavioral aspects of language are also depreciated by Deacon and Pinker but emphasized by scientists like Hall (1959) and Sampson (2005). One must keep in mind here that there are serious philosophic disagreements regarding all the versions of Causal Theory (CT) of mental content and ideas of representations (Rupert, 2001). This is due to the fact that some students of language of thought, like Cummins (1997), argue that theory mediates sensory detection. To Cummins, "CT is false in all its forms." CT has many variations. In psychology, it argues that the personality and behavior of the adult can be seen as the result of childhood experiences beginning in utero (Freud, 1966). In philosophy, Alvin Goldman (1967) summarized the idea of chains of knowledge, facts leading from one conclusion to another. Here, how terms acquire reference to events as evidence is related to Kripke's (1980) ideas of naming and linking meaning, basically a restatement of the Frege–Russell theory of referents where how speakers make reference is descriptive but attacked by Wittgenstein and Searle among others for circularity. Stampe (1977) drew a distinction between causal theories and "picture theories" of mental representations, for example, an image of (smoke) and fire or a specific event as related, or tree rings and the age of a tree. The two are related but not directly. Nevertheless, CT is at the base of empiricism—it is the means by which events are evaluated and determined to be evidence and linked together to form hypotheses and theories. The relativism of Hume and the post-modernists notwithstanding (Gellner, 1992), the world and its processes, including life, can be described and evaluated.

However, theories of the nature of language and theories of mind range from those that embrace uncertainty to those that express absolute positions. It is remarkable that Hauser et al. (2002) can speak of the discovery of the universal code of life (DNA) and the lack of a universal code of communication among animals. It seems they have forgotten that the code of DNA was only discovered some 60 years ago, and it might be a bit premature to declare that there is no universal means of communicating with animals at this point. Perhaps our means or approach has been wrong or, as yet, unequal to the task of discovery as we were regarding DNA for seven million years of hominid existence. Some experiments with teaching language to animals have touched on this problem. Deacon (1997) refers to the case of Kanzi and the success of this bonobo to learn language as an example of the critical period of language acquisition and support for a theory of the evolution of language.

The same might be true of discussions of theories of mind in other animals, since the way we have broached a theory of mind (Butler and Cotterill, 2006) has most often been based on our (culturally limited and to our species) perspective, what a mind should do, how it should function, and so on. Mostly this has been reified into goals and achievement of complexity, decision-making, planning, and deceit. Yet, other cultural perspectives (Buddhist, animist, etc.) have defined other values and purposes for mind, not encompassed in these investigations (Bateson, 1972; Spiro, 1980). Could we recognize "mind" in another species or is this concept itself species bound? Arguments of cultural brain evolution (as cultural intelligence, van Schaik et al., 2012), as opposed to the social brain hypothesis, could provide approaches here, but as yet are too limited in scope. Studies of innovation frequency and brain size in birds whether in food (Lefebvre et al., 1997) or mating complexity (Madden, 2001), though goal oriented as adaptive strategies as have been most assessments of the origin of consciousness and mind, give examples of varieties of reality assessment.

Deacon also argues that language is involved in the transmission of "meaning," though Dunbar (1996) asserts that language is often a substitute for grooming. And some theorists of discourse analysis argue that our sense of reality in everyday life and thus the meaning of everyday objects, actions and events are the product of a permanent, routinized interaction (Keller, 2011). Yet, specialists who engage in the teaching of language, especially in cross-cultural settings, find Pinker's assertions on the uniqueness of language and behavior with respect to language unpersuasive at best, and his critique of the Whorf–Sapir hypothesis unconvincing (Petro, 1996). There are also disagreements among linguists over aspects of language acquisition (Direct-Object Omissibility in English) and Pinker's (1989) argument that children are not conservative learners of argument structure (Ingham, 1993/4).

The brain-language argument of Pinker (1994) holds that human speech is unique in the animal world. He does not require a big brain for language and argues that even *Australopithecus afarensis*, approximately four million years ago, could have had speech, or even its ancestors. I will not comment in detail on this idea of unique language as I have dealt with it in depth elsewhere (Caldararo, 2014), yet while Pinker argues that animal signals (including that of bees) are very simple and an analog compared with human language, this is based on a very narrow assessment of the difference and an overblown idea of human language. It also idealizes human language function and capacity, presenting a uniform and ossified caricature. When we look at the diversity of language performance, we see that language involves many kinds of acts and inferences (Harrison, 2007), and while these signals are presented in cultural contexts learned from childhood, people respond in a variety of ways that are not uniform; there is misunderstanding, varied interpretation, and according to Fried (1975), in village life among traditional people (and it is clear also today, see Laing, 1966, for example), complete fluency was not necessary for daily life, and with the fact of marriage with spouses from different areas with varied dialects, this is magnified. And this variation in signaling to the end of comprehension is just what Chomsky (1972) was addressing as a natural history of language, and the "facts of usage" and grammar as a "natural science." For Chomsky (1972) and Pinker (1994) have argued that "virtually every sentence that a person utters or understands is a brand-new combination of words." "Virtually," I would suggest, underestimates the exceptions, for most of human communication is in distinct short sequences or repeated common colloquialisms like "How you doin'?" "How's it hangin'?" "Guten tag!" "Que tal?"

But also most tests that have been devised to assess animal language are based on human cognition and language, not just in efforts to teach animals English or Japanese. While this problem was one acknowledged by many researchers (Macphail, 1982) in the past, it seems lost on some today especially regarding imitation (Hauser et al., 2002). It often seems that the structure of testing and the underlying assumptions that result in experiment design culminate in tests for prosociality. Towner (2010) summarizes the literature and the assumptions over negative reports of imitation in chimpanzees and shows that some researchers have produced positive reports dependent on clearer examinations of the study environment and animal responses. It seems obvious in this context that other members of her group imitated the Japanese macaque, Imo's behavior (Kawai, 1965). Humans negotiate identity (gender, job, status, etc.) and use language to gain rewards and to avoid punishment. These are ontologically conditioned responses. The same is true of animals. Yet, to place an animal in a testing apparatus of human design and expect it to display abilities like normal humans is as illogical as raising a human child in an automatically cleaned room without windows with periodic food dispensers and expect it to produce language. This would seem obvious in testing abilities like phrase-structure rules (Hauser et al., 2002) where adult humans and Tamarins are compared. The experience of a human in the ontogeny of its language in a social context gives it a clear advantage over a Tamarin raised in an artificial environment.

Here is where Deacon (1997) and I agree. Language evolved in a human environment where children came to not only learn the complex signaling in a social context, but also as the brain evolves capacities for constructing maps of the environment for survival and social interaction. Selection for language was a central element, according to Deacon, where behavioral flexibility drove the general adaptive process. However, we are again left to ask why we see so little evidence of success as the brain gets bigger, and theoretically more complex facilitation behavioral variability and language? Why did hominid populations and culture not produce a more definitive social and technological complexity until after 10,000 years ago? If we agree with Deacon (1997) that language evolved in response to certain selective pressures, then one has to propose a rapid acceleration that modified it relatively recently as a precondition for human eusociality. The central problem here is the separation of language from culture, language facilitates the implementation and inheritance of behavioral flexibility. Perhaps Deacon, like many others, is making too much of language and ignoring its role in culture. Many human societies until very recently had language but their societies were not eusocial, so big brains and language do not necessarily make the superorganism context. Humans from non-eusocial societies have easily become productive members of eusocial groups, so something is missing and it is not language. In fact, both the archaeological and ethnohistorical records provide substantial evidence that language did not significantly change human economic success, that many human societies remained collectors of food, like other primates, long after possessing language.

Quine (1960) also finds that people who use the same language are never quite sure if other speakers match objects, emotions and words in exactly the way they do. Cognitive scientists often use the term "understand" as if human communication was always immediate and complete, when it is more probabilistic and fragmentary. Uncertainty abounds, and we will discuss this topic later in another chapter. Chomsky (1966) was committed, however, to the idea that language was entirely a human limited ability and has continued this position to the present day (Cucchiaro, 2007). Chomsky and his associates (Hauser et al., 2002) have denied that animal "signals" are homologues for human words. This is due to the way they have mystified and idealized human language. What are words, after all, but conditioned responses to learned contexts! They argue that the way human children "build the lexicon is so massively different from nonhuman primates," but this is not equal in all humans, and the social and economic background is essential to an elaborate vocabulary. Failure to teach animals the King's English and proper grammar after more than 100 years of trying says more of the illusions of scientists than animal capacity, especially when most humans who speak English would fail the test (Fowler, 1926; Fowler and Fowler, 1906). This problem emanates from more than just a privileged view of human consciousness, but it infects comparative language studies as when Jack Goody (1987) notes that nonliterate cultures lack words for "word" and use terms referring to speech. He seems ignorant of the fact that the derivation of the English "word" is from Old English, meaning speech, or to utter (*Merriam-Websters Dictionary*, 2017). The question to ask is not why nonhuman animals lack homologues to words (which is wrong-headed to begin with) but what does language do for humans that nonhuman animals cannot do? Language basically organizes; it is a tool for recruitment and direction of energies. Social insects also achieve this end, to ask why they do not have music, or cars or use fire, is just as irrelevant as it was in the sixteenth century for Europeans to ask why Native

Americans did not know of Christ or why some people lacked writing (of course, Europeans did not invent writing, they only learned to use it once they came under the control of more complex social regimes like the Romans). Hauser et al. (2002), like many students of language, are astounded by the variety of human words and "infinity" of ways of expression, but what value is this? It is just complex signaling; however, is more complexity better in a signal or does it promote more entropy? We might argue that language has allowed for the transition of remarkable insights into technology and economics, but to what end other than increasing the efficiency and mass of complex operations? Social insights (like those associated with the supposed "Axial Age" 2,600 BCE or so, with the interpretations of the meaning of human life as in Confucius, Buddha, and others) have not resulted in any significant change in mass human behavior since the Bronze Age. However, we will discuss this again in a later section.

Pinker and Deacon's views also assume that animal systems of communication are less plastic and have limitations to learning. Yet different bee species can learn each other's "language," and modifications in information seem quite possible, if not common (Su et al., 2008). By careful identification of individual bees and charting their behavior, Lindauer (1971) was able to demonstrate that bees could communicate to each other and modify information they had collected, as when one bee returning to the hive in search for a new hive location met another with a different location (in mind), then journeyed to the second location for comparison, then back to the hive, and then, incorporated the second location instead of the first. Such detailed analysis of communication, editing, and evaluation should be easier to accomplish today with contemporary computers than in Lindauer's day, and more recent reports support his observations (Su et al., 2008), though the fact that a tiny animal with such a small brain can accomplish such complex perceptual assessments escapes explanation. This process has been reported with other social insects (Visscher, 2007).

Visscher and Seeley (1982) proposed that dance dialects evolve as foraging maps (as did Gould, 1982) and their variation may indicate links between ecological adaptation and the creation of distance codes in the bee language, though Dyer and Seeley (1991) report some contradictory evidence. This could be another element in the ecological theory of brain evolution. Therefore, if we join the ecological theory with the social theory and add the idea that language also evolved parallel to other means of integrating elements of the environment, as in tool making (Schick and Toth, 1993; Wynn and McGrew, 1989), it becomes significant since we know that the same areas of the brain that are activated in speech (e.g., Broca's area) are also so in tool making and music (Arbib, 2012; Stout et al., 2008), other suggestions by Lieberman (1984) and Bickerton (1990). Thus, extended fields of action and interaction in the brain can be seen as expressions of mind, with interlocking selective advantages. Recent studies with fMRI have identified a "semantic system" of a group of regions in the brain with areas selective for specific semantic domains (Huth et al., 2016). Further research is necessary to define and describe the extent and formation of these regions and their interaction. But it seems that the area where this can be interpreted as "words" is that of neuronal relations produced through complex systems of conditioning.

In a different use of fMRI, the anterior cingulate has been associated with cognitive conflict involving strategizing in game theory studies (Camerer, 2003). And where categorical speech perception has been studied using similar techniques (Bidelman et al., 2013), neurophysical underpinnings of transformations in the brainstem and activity in the auditory cortex provide potential representations of acoustic to phonetic mapping. This may reflect the kind of physiological distinctions in speech noted by Liberman in 1957; it has however also been associated with visual stimuli that release specific behaviors in certain innate systems in animals and plants. This falls, regarding speech, into the category of speech perception and categorical perception required for efficient transmission of ideas (Holt and Lotto, 2010). Perceiving or constructing specific categories from spectrums is an essential element of responding successfully to the environment of noise in the evolution of specific organisms' adaptation. Evidence for neural mechanisms for lexical processing in dogs has been reported (Andics et al., 2016) using similar techniques and fMRI. Yet, organization of neural representations may take place in the human posterior superior temporal gyrus (Chang et al., 2010). In humans, and perhaps other animals (Fisher, 2006), it appears that such categories may be platforms of learning, as in color perception or language. Gene variants in color vision in humans make this even more complex (Verrelli and Tishkoff, 2004).

Regarding the discussion above, to be fair to Deacon, however, we must note that most of the work on animal cognition that has changed our views has been accomplished only in the three decades since the publication of his book. I have used his book as a foil only because he presents in it the general "human unique" argument that is still so convincingly prevalent today.

Deacon (1997) criticizes the idea of animal communication and considers it protolanguage, and I think along with Striedter (2005) that this can be an example of *scala natura*, where animal communication is seen as primitive language. However, if one considers animal communication as unique evolutionary sequences of adaptation, then the idea of protolanguage depends on our ability to discern its existence and complexity. Humans did not realize that bees had a language (the ancients had a general idea), but it took a scientist, von Frisch, applying the correct questions and, using precise methods, to discover its complexity. This process of discovery is continuing today with other species (de Waal, 2016).

Human language, it seems reasonable to assert, is no more than a complex system of conditioning developed from the context wherein human babies are born with only about 27 percent of their brains developed (more than some rodents, but less than other primates) and the brain develops within a social context where vocalizations become stereotyped into what we call concepts, organized in segments of communication that we refer to as grammar. Ideas about what "developed" means vary, but functionality is essential as, for example, the increase in synaptic density and dendritic arborization at about 7–10 months that coincides with a rapid improvement in working memory (Semple et al., 2013). The human neonatal is born with about 27 percent of the adult brain mass (Blinkov and Glezer, 1968; DeSilva and Lesnik, 2008). Their findings contrast with Martin (1983) and do indicate a smaller human brain at birth than expected of adult brain size. Not all that special, in some ways: rat and mouse neonates have about 18 percent (Nieuwenhuys et al., 1998), while the Weddell seal neonatal is born with the largest brain of any mammal at 70 percent (Eisert et al., 2013). We do know, however, that other animals with complex societies, bees, for example, have longer life spans than solitary or non-hive bees (von Frisch, 1953, 1967, 1971). We might assume, therefore, that the evolution of complex society produces many traits, such as longer life span, that we associate with human uniqueness.

In studies of human evolution and the evolution of the brain, especially biocultural views as well as those derived from sociobiology, researchers have attempted to contrast the evolution of human consciousness from a number of theoretical directions. The next sections provide an overview of these approaches.

A. EXTRAPOLATING FROM THE ARCHAEOLOGICAL RECORD

One approach proceeds from examining the remnants of human anatomy and environmental exploitation, that is, the fossils, stone tools, campsites, and debris found in the archaeological record (Schick and Toth, 1993, but contrast with Bunn et al., 1986, and Binford, 2002). Isaac (1979) argues that six hypotheses of human evolution can be examined under this heading: Dart's (1949, 1954) hunting hypothesis; Jolly's (1970) seed-eating hypothesis; Tanner and Zihlman's (1976) gathering hypothesis; Isaac's (1978) own food-sharing hypothesis; Parker and Gibson's (1979) developmental hypothesis, and Lovejoy's (1981) shortened birth hypothesis. Alexander (1989) rejects all of these, and restates and elaborates on a hypothesis suggested by Humphrey

(1976), which is very similar to that promoted by Dunbar (1992, 1993, 1998) except that he adds the idea of the predator behavior as significant, but no predators except wasps and ant species have formed complex society (social Synalpheus of the shrimp family is considered to be eusocial, but is a parasite on the sponge host [Duffy, 2002; Duffy and McDonald, 1999]). Complex society here refers to definitions of eusociality. Brain size and social organization among predators do not support this idea (Finarelli and Flynn, 2009).

Alexander (1989) also argues that a process of "runaway competition" is central to this evolution. But one might argue that a more complex process can be imagined, one where the creation of rules of order within and between groups, based on core and range interactions in other primates, but expanded by the interaction needs of providing defense and environmental exploitation systems drives complexity in brain evolution and rewards it in selective pressure. Here, the cosmological construction we see in religion not only functions to maintain group integration (increasing group fitness), as Durkheim (1933) described, but also as reference to data on people's rules of life and conception diversity as in Lee's (1959) or Hoebel's (1964) demonstration in law, or, for that matter, law in general and how people come to obey and respond to institutions (Foucault, 1977; Hodgson, 2006).

It is not either/or between competition and cooperation, but a balance related to the environmental pressures that keeps them in check by rules and institutions. Yet it seems to me that Alexander (1989) overstates the "runaway" aspect of competition among hunter and gatherers, for even among a group like the Siriono (Holmberg, 1969), who were under severe stress and continually hungry and living in distress, there was a delicate balance of intergroup relations. But this merges into our next thesis because the rules demand that humans know how to interact with all life, to understand the movements and intentions of other animals, and often, the aspect of groups of life as in forests as consciousness to be "understood." The ecological intelligence hypothesis (EIH) (Steklis, 1999) fits into this group, and Dunbar often joins the social brain hypothesis and the EIH. A variation of this concept is the cognitive-buffer hypothesis of large brain evolution. The central idea here is to balance the costs of big brains by an enhanced ability to handle novel situations and hence with increased probability of survival in novel or altered environments (Sol, 2009).

One can argue that the inventions of the Japanese macaque, Imo, on Kashima Island, are an effective example of this theory (Kawai, 1965). I am inclined, however, to consider Imo's actions to be like other unique interventions that animals (and some humans) produce. I call this attentive presence, where actors are able to penetrate their conditioning to reformulate material or social aspects of their environment (unlike like the conscious awareness of Griffin, 1976). In Imo's case, she invents potato and rice washing. This kind of intervention seems independent of specific physiological cognitive bases and thus might be considered a cross-species ability appearing in only certain individuals. Of course, it might be a sequela of disease or trauma. For example, one might class the remarkable insights of the neurologist Oliver Sacks (1970), a man who could not recognize faces of patients (the condition of prosopagnosia), with the perceptions of a man who had no memory and no past.

Peripheral to this, but related to ecological demands, we find the work of researchers who link the evolution of mind as a collection of modules. This is characteristic of the research strategies of many evolutionary psychologists following on arguments by a philosopher, Jerry Fodor (1983). The idea was first proposed by Flourens in the nineteenth century (Pearce, 2009) and has been taken up by Pinker (here, he presents the idea that specific functions of mind, like perceptions and language, are associated as organs into modules of processing yet are independent of each other). Karmiloff-Smith (1996) has shown that the modular approach is not a robust explanation for what we find in developmental problems and degenerative conditions (her area of research is Williams syndrome). Her research and that of many of her associates makes the modular concept of mind (and other domain-specific theories) rather untenable. Her book, *Beyond Modularity* (1992), and Jeffrey Elman's *Rethinking Innateness* (1996) have been devastating critiques. While the evidence for the modular hypothesis seems lacking, neurophysiologists like Striedter (2005) are more concerned with connectivity and the integration of regions of the brain as the result of two different theoretical approaches: parcellation (associated with Deacon's rule) versus the invasion hypothesis (Striedter, 2005).

B. COMPARISONS WITH SURVIVING "TRADITIONAL PEOPLES"

Known as the Direct Historical Method (pioneered by Cyrus Thomas, 1898), this approach can be traced to Worsaae in 1843, as noted by Rowe (1962) and later developed by Strong (1935). Holloway (1975) elaborated on this. Here also, we find ideas of differences in mind produced by life-ways or culture as described by Levi-Bruhl (1923) and modified by Jung (Segal, 2007). This reevaluation has similarities to psychological nativism linked to the work of Chomsky and Fodor (Ferretti and Fodor, 2001). Some evolutionary psychologists have expanded this view to an adaptationist process where functional partition of the brain is a result of the hominid evolutionary past, as a result of hunting and gathering (Duchaine et al., 2001). Central to their argument is the existence of a computational adaptation, but as described, and in definitions of such processes, logic and form of reason are required. The problem here is that the desire for logic and the expectation of reason are cultural traits and organized differently in various times and places. As S.C. Levinson noted in a 1995 paper, there appears to be a social bias in all thinking. Robin Fox argued, in a 1992 study that "rationality cannot be equated with logic." From reviewing the results of a number of studies in cognitive science, H. Dieter Steklis (1999) found that there appears to be at least two kinds of rational thinking, and regarding S.C. Levinson's work, suggested that social cognition steers all thought toward illogic. We have to be careful in the use of the term "culture" as the transmission of learned behavior (knowledge) from one generation to the next, as some insects, for example, ants do transmit such adaptive information generationally (Rosengren and Fortelius, 1986).

C. USING SURVIVING PRIMATES AS MODELS

Another approach is to contrast human behavior with surviving primates as models of earlier stages of human behavior, especially the apes (see Stanford, 1996 for an example and Jolly's 1970 hypothesis could also fit here). A different theory of the role of sexual selection proposed by Parker (1987) would also relate to this approach as would the one by Tanner and Zihlman (1976) extend the effects of behavior to the results of innovation and selection. This is similar to Dunbar's (2010) concept of group size as a factor in brain evolution. Yet, examples with bower birds (e.g., Ptilonorhynchidae) show increased brain size in species that build complex decorated bowers; thus, the social aspect seems small except for female choice (Madden, 2001). The complexity of bower, however, appears to be taught to immature birds. Innovation and novel designs are significant factors in success, as is spatial memory for locating and maintaining caches of materials.

D. IDEAS OF MIND

This approach employs theoretical propositions of consciousness function, explained through physics, biochemistry and philosophy (Broadhurst et al., 1998; Horgan, 1994). Under this category would fall the idea of the human brain as a by-product of specific adaptations of the body, as in Fialkowski's (1986) theory of thermohomeostasis and hunting, later elaborated on by Falk (1990), or the pressure of disease (Caldararo, 1996) and the avoidance of disease (Caldararo, 2012b; Loehle, 1995) as well as the similarity of neurons and immune cells to transmit signs and information about specific environmental states to each other, and the complexity of that information (Dissing-Olessen et al., 2014; Ferber, 2007).

We can place ideas of mind as a collection created by its parts, a unity or as Lashley (1929) put it, a form of mass action containing equipotentiality, allowing the brain as a whole to compensate, in certain cases, for the loss of portions and has been opposed to ideas of strict localization. This concept was derived from experiments where parts of the brains of animals were removed or destroyed, and the overall effect on the performance of the animal assessed (Lashley, 1929; Tsang, 1934, 1936; Zangwill, 1961). Here, one can contrast the idea of a brain versus that of processing information and simply "mind" or structure of organ versus function. Reber (2016) takes perhaps the extreme position that "All organisms that experience have minds, all have consciousness."

E. ANATOMICAL COMPARISONS WITH BRAINS OF OTHER ANIMALS

In the 1960s, Lilly (1961) began to question the comparison as based mainly on a hierarchical system with the human brain at the apex; everything was thus in contrast. One might place Parker and Gibson (1979) here, but the book by Parker and Mckinney (1999) is a more comprehensive discussion of the ideas of development, especially that of "adultification by terminal extension," which falls under a number of scenarios discussed by Striedter (2005). They also deal with theories of extended childhood and neoteny. Attempts to adjust this bias by Jerison (1973) and others (Count, 1947) failed to produce a reliable measure of brain size as a comparative metric due to problems with the design of variables, for example, brain to body mass, or surface area ratio or regional area proportion (Striedter, 2005). Preuss (2001) placed emphasis in this regard on the reorganization of the human visual system and Falk (2007a), citing work by Finlay and Darlington (1995) of the significance of regional size to overall brain size. Striedter (2005) notes that in looking at the function and brain size in relation to regional proportions, reference to studies with corvids and other birds that cache food gives a potential adaptive relation between the behavior of caching and the relative size of the hippocampus where enhancement of spatial memory would produce selective advantages.

The main problem with this is not only the lack of correspondence across species, but also the problem of overall effect of behavior in absolute brain size and regional proportional development. Still, this relationship of larger hippocampus and food storing is also found to be related to complex concept formation in blue jays (corvids, see Healy and Krebs, 1992) as Macphail (1982) has noted, in tasks like learning set formation that parallel abilities in primates. However, just as later fossil evidence shifted focus to specific brain convolutions and the significance of asymmetry in the 1990s (Holloway, 1996), concerns of conductivity became prominent in the past decade. More recent reviews place emphasis on convolutions and connectivity (Hofman, 2014). Here some have sought complexity as a representation of progressive evolution, yet this may be illusory as Striedter (2005) notes in the similar pursuit of brain complexity by accretion as opposed to phylogenetic segregation. This is particularly frustrating when it is noted that brain complexity has increased and decreased several times in evolution. And it is difficult to measure complexity as a function of increased numbers of brain regions, under the assumption that more regions mean more complexity, as if simply more feet or stomachs indicated progressive evolution.

Relating brain structure to function is problematic, as assessment of outcomes is often fraught with subjectivity due to the nature of the subject. One can imagine how the choice of testing subject can affect outcome. Consider that if space aliens chose macaques to represent primates due to the ease of handling this species, their assessment of the Order might be significantly different than if they had made a less expedient choice, that is, humans. Macphail (1982) demonstrates the importance of contextual variables in tests, but to create a comparable context for humans and other vertebrates, one would have to seize humans from the street, inject them into a room and nonverbally direct their behavior through a testing protocol. Deaner et al. (2006) examined primate cognition studies to determine variations in testing design and animal response; like Macphail's (1982) review, it provides a basis for caution. This is also Roeder's (1970) concern regarding surgical interventions. Kraus et al. (2014) have called into question much of the testing results with vertebrates due to the testing design handicapping many species. Tomasello et al. (2005) can speak of children of 14 months "understanding" adult intentionality, producing "rational imitation," but using these terms distorts the behavior observed (Herschensohn, 2007) as using similar terms for birds is considered anthropomorphic or projection. Yet, this ascription of intentionality is assumed and cannot be measured empirically as Perner and Doherty (2005) have pointed out. Other workers have noted problems with this approach from different animal research including chimpanzees, lions, hyenas, and dolphins (Hatano and Takahashi, 2005; Horner et al., 2005; Kuczaj and Highfill, 2005; Schuster, 2005). It is interesting, however, that tests with insects have shown increased abilities on a number of tasks related to complexity of structures like mushroom bodies, where some insects can be considered as "sophisticated as some mammals" (Burns et al., 2011). This development of the mushroom bodies and related structures is greatest in insects that forage widely and hunt, especially returning to certain areas and revisiting cached resources.

In this is a parallel with humans, as well as certain birds and cephalopods (Farris, 2013).

What is remarkable in the Tomasello et al. (2005) article is a reversal of earlier views that representation of mental states was unique to humans, after demonstrating that they could find it in other primates, and then, in this work that it is intentionality that makes our cultural sophistication possible. The continuing problem here is the veracity of the "demonstration." Motivation is, in their assessment, lacking and yet this enters into the mind of individual animals, a quite difficult arena to test and explain (Lyons et al., 2005). Ideas of mind and self are often confused when discussing human action as in "cognitive structure," "schema," or "hypothesis" (Kagan, 1970) or Freud's ego, superego, and id (1966). fMRI images can provide us with physiological notation of brain activity (Andics et al., 2014), yet translating this action into ideas of mind or self is difficult to achieve without projection. Does such similar intentionality mean we can say what the blue jay is "thinking?" Demonstrations of loss of function, as in aphasia due to structural damage (Geschwind, 1970), allow us to conceive of how sections of the brain work, but to identify loss of function with loss of mind is subjective. Macphail (1982) suggested that experiments with nonhuman vertebrates seemed to indicate that "humans are more intelligent simply because they possess language"—language being, in this view, a more effective mechanism for thinking. Yet as we will see below, the fact that nanocephalic dwarves can learn and produce language would seem to undermine this theory. There is evidence of an evolutionary relationship between thinking and language, for example, in the execution/observation matching system found in monkeys and humans (Binkofski and Buccino, 2006). It was thought that the granular prefrontal cortex, where "rational" aspects of decision-making could be associated (Striedter, 2005), was unique to primates (Preuss, 1995), but work by Uylings et al. (2003) with respect to decision-making has found evidence of this organ in rats. This only demonstrates that increasingly, as our techniques improve, we are finding that what was once thought to be unique to humans is shared with other animals.

While the experimental design can have a significant effect on the assessment of a subject animal or human's ability, the attitude of the experimenter toward the internal life of the subject can also skew the interpretation of the nature of mind and behavior. For example, in Koehler's (1950) experiments with birds concerning the subject's recognition of number concepts, he found that ravens and parrots could choose from a collection of five boxes having lids with different numbers of spots on them. Koehler did not claim that the birds "counted," and rejected the idea that they used "words" because he did not believe they could "name" things. Here again, there is little evidence to support Koehler's assumptions that the birds could not count, name things, or "know" anything like what humans do. Macphail (1982) questions Koehler's attitude citing other work by some of his associates. Recent support for Macphail's position has come from studies of New Zealand robins (Garland and Low, 2014).

The theoretical framework of the experimenter is central to our understanding of the nature of animal behavior and the discovery of the nature of mind. Skinner's (1948) ability to produce in pigeons' behavior very similar to what can be called "superstition" was convincing, but not if one considers the irrationality typical of people's economic behavior, which is supposed to be the ultimate in the rationality that defines human mind (Caldararo, 2009c; Etzioni, 1988; Vyse, 1997).

In general, this discussion takes place on a plane of isolated assumptions, for example, all the people have always used languages like those existing in most parts of the world today, ignoring sign language and "click" languages. The implication being that one has to have verbal language spoken in certain ways to be human excluding people who are unable to "speak" or those who use minority forms of speaking. The evolution of language is usually missing from the discussion, but then since we have no means of analysis of this process, no archaeology of ancient protolanguages, this seems impossible. Pinker (1994) demonstrates how plastic the acquisition of language is among children and how they are able to take combined systems like a pidgin and "inject grammatical complexity" where none existed before. In fact, Pinker argues that children's ability to correct adult problems in teaching language, as in deaf children, demonstrates a unique capacity for language, yet he never asks why so many adults today fail to acquire proper grammar themselves. Griffin (1976) addresses teaching across the animal kingdom, as does Bonner (1980), but Premack (2007) asserts that teaching (as in other crucial behaviors) "did not evolve in the context of food seeking" as in other animals, but in a broader context of language and aesthetics. Unfortunately, this cannot be demonstrated, as we cannot determine the veracity by traveling back in time to analyze the context of the evolution of human teaching; but certainly the context of the evolution of tool making did take place, and most likely, the teaching of tool making in the context of food seeking, as we see with chimpanzees (Schick and Toth, 1993).

Deacon (1997) argues that abandoning the search for antecedents by positing an instinctual basis for language (generative grammar, etc.) is illogical. He also makes language a driver of brain evolution. But a central question to his idea that language and the "mentality" necessary for it is unique to humans is how would one tell if other species had language? This is the question Sagan and Shklovskiĭ (1966) ask of how to communicate with aliens. How does one recognize other systems of language if language is species specific?

Pattern recognition associated with behavioral responses is one proposal Mat Cartmill presented some years ago (1999), but this implies an understanding of other being motivations and intentionality, something humans have been poor at even among themselves (has led to wars, civil conflicts, and divorce). Laing (1960) attributed this to the elements of individual psyche itself and the nature of phenomenological uncertainty.

Cartmill (1990) questions the paradigm most often used in applying the idea of language to animal communication experiments. But Deacon (1997) sets up an inequality by using the term "non-linguistic communication" for expressions like laughter, sobbing, hugs and kisses, and argues that these expressions produce instinctual responses, as when someone enters a room laughing, he asserts everyone laughs as a result. The fact that this cannot be supported by scientific data (Smith, 2009) is a problem, as is his juxtaposition of "calls" to "words" as if we knew that "words" had no history, which implies that they did not evolve and are instinctual. But, he uses the idea of things becoming "icons" or "index" for meaning or symbols. Yet, this is not an explanation.

He also uses definitions to make conclusions about language and nonlanguage as when he states, "Alarm calls refer to objects the way laughter does, not the way words do." However, no one would deny that the word "stop" is an alarm call. As mentioned above, this is partly due to Deacon's (1997) de-emphasis of the role of behavior in language, but while Deacon allows gesture and attitude to have a role in animal communication, he downplays it in human linguistic transmission, a conclusion in contradiction to the work of Hall (1969) and Hymes (1996) in anthropological linguistics. As Griffin (1976) states, every list of requirements for language is defined by human language, and yet even here we can find animal examples, as for instance, the bee waggle dance is surely an "icon" for the route to the food source. More to the point, prairie dogs produce novel alarm calls in response to unique intruder appearance. This can only be a process like the production of representations of that appearance, or symbols, as sound units (Slobodchikoff et al., 2009).

Deacon (1997) makes a common projection on the concept of humor by arguing that while chimpanzees make a sound similar to human laughter "they would likely miss the reference to humor." As he believes that laughter equals humor and a requirement for it is contradiction or paradox, certainly an ethnocentric view of both laughter and humor, this leads him to another projection. He assumes that chimpanzees are not capable of recognizing either contradiction or humor, but one has to ask, how would he know? Some researchers have found behavior like human laughter in birds, for example, and associate the emotion in animals with an early evolutionary expression of joy (Panksepp, 2000). So, while Deacon may believe that humans (really westerners and perhaps mainly Anglo-Americans) laugh due to contradiction or paradox, to "laugh" might be a more complex response in vertebrates.

While behavioral responses similar to sadness, depression and laughter can be elicited in other vertebrates by brain stimulation techniques (Panksepp, 2007), human researchers must interpret these behaviors into culturally determined categories. Work with apes has also demonstrated a variety of responses that appear similar to human laughter in different contexts (Ross et al., 2009). We must be careful when evaluating what we can do and comparing it to what other animals do given not only the difficulty of interpreting the responses of other vertebrates, but the fact that we are often unaware of their particularly formed response, as in the case of recently discovered hearing range of a form in Borneo, *Huia cavitympanum*, that hears in the 20 kilohertz range. Often, our ability to sense the response of an animal is as limited or distorted as our competence to interpret it.

Deacon (1997) asserts, "The human brain should reflect language in its architecture the way birds reflect the aerodynamics of flight in the shape and movements of their wings." Perhaps it is better to use the same metaphor to make the point: the human brain evolution is to language what tool making is to other forms of food collection as the hand is to the paw or beak. Though I would not agree in either case. While this is a poetic view, perhaps a more fitting analogy would be the differences between how mammals process visible light, and how birds process and integrate visible light and UV.

ORIGINS, ANTECEDENTS, AND PURPOSES

These five general areas of study have produced a variety of theories, as I have noted, yet language is most often the behavior that one finds central to the discussion of the unique behavior of the human brain and expression of mind or consciousness. Therefore, an examination of language seems a logical next step. Why language exists seems simple: to recruit members, to organize their efforts and to mediate their needs both physiological and psychological. Dunbar (1996) placed grooming—found in most primates as a means of social reinforcement, vermin removal, and status display—at the foundation of language. Grooming is found in many animals from cockroaches to ants, and functions to clear debris from antenna and reduce potential disease and infection (Zhukovskaya et al., 2013) to extension to the social context of the nest (Caldararo, 2015a; Rothenbuhler, 1964). This does not always succeed, and eusocial colonies appear to be more susceptible to infection and pandemics than less integrated colonies or solitary insects in some cases (Tragust et al., 2015). However, when social immunity is present in eusocial insects, the opportunity for networks of cooperating nests develops and this increases genetic diversity creating a resistant and sustainable form of superorganism (Ugelvig and Cremer, 2012).

Humans also suffer a complex "symbiogenesis," to use Margulis's term for the occupation of one organism by a number of others (Margulis and Mikhaylovich Kozo-Polyansky, 2010; Margulis and Sagan, 1995). How long we have possessed intestinal bacteria and other symbionts (as well as endogenous viruses) is unknown, but some endogenous viruses indicate histories of millions of years (Caldararo, 1996), and intestinal and other symbionts are recorded from archaeological and DNA studies (Sender et al., 2016). Yet, it seems that social insects in unicolonial systems have arrived at a similar plane of development as contemporary humans. Not only is hygiene highly developed, but immunity is transferred on a systematic basis and genetic heterogeneity maintained by different colony inclusion (Ugelvig and Cremer, 2012). The balance of conflict is not between social insects of the same species nor between the species of the same or different genera or orders, but between their parasites and pathogens for the survival of social complexity (Caldararo, 2015a). Humans had an added element of support in this struggle in vaccination and antibiotics, but in the race for profit and mass production of vertebrate food sources, pathogen evolution has surmounted the present capacity for control (antibiotic resistance). For both social insects and humans, the limits of complexity may have been reached.

Language must have had antecedents, and even if it were the result of a mutation as Richard Klein and Blake Edgar (2002) argue, claiming that a rapid increase in material culture can only be explained by a unique and immediate change in the genetics that regulate the brain's chemistry. This is in some ways a restatement of Pfeiffer's (1982) idea of a creative revolution in the Upper Paleolithic (itself a restatement of Krantz's [1961] concept of some brain change), a concept that is undermined convincingly by the findings of McBrearty and Brooks (2000). While efforts had been made to discern more complex behavior earlier by methods that could discover residues of complexity (Fuentes, 2015), the results have been unconvincing (Brodie, 2005; Dawkins, 2004), though I find niche construction theory (NCT) appealing as a potential source of information (Laland and O'Brien, 2010). It has a parallel in Dawkins' (1982) idea of the extended phenotype, as where a beaver dam extends the ability of the beaver to exploit and manage its environment, or the spider its web, in much the same way tool making and language do for humans. This extension includes the organisms that humans and social insects domesticate, and the coevolution of this relationship is a significant aspect of the history of multilevel selection in the eusocial systems as Goodnight (2016) has emphasized.

The work in niche construction theory gives credence to other similar efforts to establish traces of human activity and a great phenomenon of the extended phenotype, spirituality. Roheim (1945) and Mullen (1981) as well as the more recent efforts of Geoffrey Miller (2000) have placed emphasis on the relationships of sexuality and religion as a means of understanding the evolution of the human mind. It is obvious that spirituality does not necessarily leave any material traces, yet the creation of cosmologies, both naturalistic and spiritualistic, is a complex system of thought and requires the engagement of others in learning, convincing, recruitment, and conversion. The last of these is, perhaps the least understood today (Barro et al., 2010). It is of interest, however, that stages of religious conversion and those of schizophrenia are quite similar. Wooton and Allen (1983) found that

> [t]he similarities between dramatic religious conversion and decompensation to schizophrenic psychosis are evident from the review of the literature. Each of the two phenomena proceeds in a series of characteristically identifiable stages; the initiating condition or conditions of the two processes may be similar or even identical, and the stages in the process of religious conversion correspond to the early stages of decompensation.

The central difference was that the convert came out of the process with an intact ego. Such a process of reformulating "faith" may involve accessing and modifying "cognized" knowledge, a form of unconscious information Chomsky (1980) has described. A substantial problem with this view is the lack of uniformity in definitions of conversion and schizophrenia (Nissimi, 2004; Rambo, 1993).

We have to recall that the human brain size had reached a maximum range at about 150,000 BP and Klein's estimation of the culture change took place 100,000 years later. This idea is curious to me as it infers that every time there is a significant change in material culture, it means there has been a change in the genetics of brain chemistry; such an idea seems forced. Does that mean that the end of feudalism and its low productivity as well as the appearance of a more ordered and productive Renaissance, the Enlightenment and the Industrial Revolution indicate a change in brain chemistry? Could the post–World War II tech revolution leading to the resistor and Apple computers indicate the same? This appears to be the thesis of Cochran and Harpending (2009), but they begin the period of rapid evolutionary change at about 10,000 years ago, and so we cannot include them in the idea of rapid "hominidization" (Tobias, 1971).

In discussing niche construction, we must also consider the complexity of the animal's intervention in the natural environment; thus the social insects have achieved considerable levels of intervention, but so have some mammals, such as beavers (dams and living spaces), prairie dogs, and meerkats. Dawkins' (1982) addresses this briefly for beavers in his "extended phenotype" idea. Mitochondrial DNA indicates that contemporary dam-building beavers arose in the early Eocene (Horn et al., 2011; see figure 2.1). The ability of beavers, members of the order Rodentia, to alter the landscape has been recognized since antiquity, yet the elimination of their habitat and the shrinkage of their range with human activity has resulted also in the destruction of most evidence of their structures, and the social groups necessary to maintain them. In May of 2011, satellite images showed the largest beaver dam in existence in Canada at 850 meters (http://www.cbc.ca/news/technology/alberta-s-huge-beaver-dam-sparks-media-buzz-1.890756). In Castor fiber, at least, instinct may play a significant part in initiating dam building, as Wilsson (1968) found that a Castor fiber infant raised by humans began to build a dam on hearing running water. Though anecdotal evidence, it may have some relevance. However, Damilov and Fyodorov (2015) argue that dam-building is a response to specific features of the environment of beavers. This is consonant with Wilsson's (1971) finding that dam-building is influenced by learning.

The other theories of rapid "hominization," as compelling as they might be in a romantic sense, are not supported by fossil, biochemical or genetic (or as McBreaty and Brooks [2000] show, technical) evidence. If we investigate this general hypothesis of rapid "hominization," and by means of it, also Klein's,

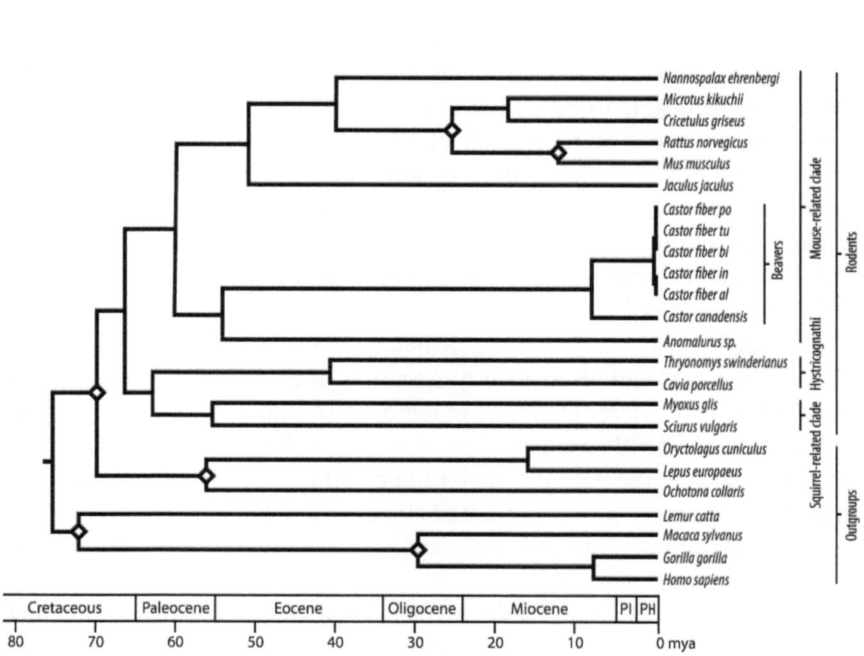

Figure 2.1 Chart of mammalian evolution. ***Source*****: Redrawn from Horn et al. (2011). Used with permission.**

we have to accept the fact that human societies before 10,000 years ago lived lives of nomads. They were hunters and gatherers, not food producers, and their behavior differed little in that regard from other mammals. While bees and some species of ants were food producers (ants were both gardeners and domesticators of other insects perhaps as early as the late Cretaceous (Schultz and Brady, 2008), humans did not develop domestication of animals until after 10,000 BCE. Evidence of this process shows a slow development of adaptation from gatherer and hunter to full domestication. Sedentary life, or at least partial sedentism appeared after this date, but the complexity of that transition and the intensification of domestication of animals and plants to provide full reliance was uneven, and we might argue conservatively that it was well established by about 6,000 or 7,000 BCE. We should be careful in this assessment, as we have examples of eusociality in human groups without agriculture especially in the Pacific Northwest, for example, the Kwakwa̱ka̱'wakw (Kawkiutl). The enormous, plentiful resources promoted collective life and foraging with a dense population and institutions for redistribution, power, and prestige (Boas, 1925) that was still subject to seasonal variations causing their urban sites to be partially depopulated at times. Even reports at contact promote a view of eusocial life (Strange, 1785–1787).

We should consider, also, that the first sedentary communities are evidence of a marked change in lifestyle and that the accumulation of human waste and dense conditions likely promoted new morbidity rates. Former low hunter-gatherer birth rates, as adaptations to carrying capacity of the land, would be unable to match losses due to increased disease burden. This would require a loss or repression of earlier forms of population control (e.g., infanticide), yet under these conditions, more infants would die as more were born. Therefore, we have the explanation for agricultural fertility increase that set up a continuous cycle of early death, increased birth rates, food production, increased population, and increased death rates (Fabrega, 1997; Greaves, 2000).

Foley (2001) commented on the difficulty to identify human social institutions in the archaeological record, assuming them to exist in some form before 50,000 BP. Here, Dunbar's argument (de Ruiter et al., 2011) would seem to apply as his theory posits an association of cortical size with group size, and relates to the needs of social life creating pressures in selection for larger brains to cope with group interactions and individual survival within the group dynamics. This may only apply to vertebrates, as group size in social insects and brain size varies, but we will discuss insect brains later. While Dunbar makes a convincing mathematical argument among a number of vertebrates (though actually the central association is not group size but increased body size across genera, see Striedter [2005]), the problem with the theory is why so much brain size was necessary to explain the specific needs of intragroup or intergroup selection in hominids and only after 2.5 ma, but then decreasing after about 50,000 years ago. The chart he produces is not based on strict scientific testing as de Ruiter et al. (2011) have noted, and utilizes a broad number of assumptions that have long been rejected in research on animal ability (Macphail, 1982), that is, there is no rigorous definition of situations, communication and relationships, simply subject conclusions.

More troubling is the lack of specificity in comparisons with other vertebrates on brain size, energetics, life history patterns, and ecology (Harvey and Bennett, 1983). Applying these factors in an allometric relationship across 15 orders of mammals, Harvey and Bennett (1983) show a much different distribution than usually seen in charts from Jerison's (1973) computations given brain-to-body ratio and his scaling exponent. In general, small animals have larger brain mass to body weight (small rodents 10 percent, man 2 percent, pig less than 0.1 percent, blue whale 0.01 percent), but most scientists studying brain size across genera attempt to control for body size by using allometry, producing a variety of equations to achieve a relative sum, or encephalization quotient (EQ) (van Dongen, 1998); however, differences in large and small animals can be considerable empirically producing slopes of 0.67–0.78 (Jerison, 1985). Boddy et al. (2012) have calculated EQ for 630 mammals and applied a number of tests to the data, concluding that 0.75 had the best fit and that there are considerable variances in EQ within mammalian lineages, with humans and dolphins having the highest EQs.

A cultural deficiency appears in Dunbar's idea, and one that is common regarding nonurban and nontraditional people; he assumes that people in the past have been frozen in place and developed their relationships only from those nearby at hand. We know this is not true from extreme examples of relationships at distance as in the Kula Ring partners (Malinowski, 1922), but again Dunbar's number depends on assumptions: what "stable group" means and how technology affects this over time (Krotoski, 2010), or measurement difficulties of Chapple and Arensberg (1940). It also reminds one of other attempts to achieve the same end, as in "Miller's Law" published in 1956 by cognitive psychologist George A. Miller as the limit of one-dimensional absolute judgment and short-term memory, though Miller recognized that there were limits and used it rhetorically. It is close to the phenomenon discovered by Crowder (1976), where the limit of recall in humans was found to be seven, yet affected by recall with short-term memory capacity at two to four and recall having serial position effects. Nevertheless, it has been followed by ideas of limits to working memory capacity, as in Baddeley (1992), Shiffrin and Nosofsky (1994), and Cowan (2001) with "chunking" of information. Yet, Zhang et al. (2005) report on experiments that demonstrate not only working memory in bees, but "symbolic" rules for navigating, responses that in a number of contexts challenge the idea that symbolic behavior and abstraction are limited to humans. Experiments with pigeons (Boakes and Gaertner, 1977) appeared to demonstrate symbolic communication between two pigeons, but they surmised that it was the result of "autoshaping," certainly a complex aspect of awareness where one animal notices the sequence of behavior and rewards in another and follows them.

One major contribution of Striedter (2005) was his criticism of the scala naturae concept in biology, mentioned above. This idea of increasing complexity to the final product of humanity has distorted our ideas of evolution and rendered the surviving members of different earlier forms (e.g., fish) as frozen representatives of an evolutionary sequence leading to primates and ultimately, man. Yet, we know from the comprehensive study of plant and animal life that all forms have evolved and that the survivors of orders that appeared earlier in time have themselves, in most cases, produced significant complexity both in structure and behavior. Macphail (1982) exhaustively demonstrates how the history of experimentation often structured our understanding of the evolution of behavior along the concept of scala naturae due to both the preconceived ideas of the experimenters, the structure of experiments, and the lack of diversity in species utilized. For example, he shows that almost all behavioral experiments with fish were conducted with teleosts by the late 1980s. Today, we benefit from a mass of experimentation that responded to Macphail's analysis, as well as new instrumentation, and a greater diversity of species under experimentation that has found not only a significant variety of behavioral complexity in many animal orders than previously thought possible, but in some behavioral components once believed to be limited to more "evolved" orders, such as Machiavellian intelligence in fish (Brown, 2014).

Here Steklis (1999) brings up the point of social bias in all thinking discussed by Levinson (1995) and Goody (1995). They argue that social intelligence is not logical, yet Steklis (1999) derives from their positions that there are several kinds of logic and he suggests that the origins of rational thinking and logic must have been established in adaptive problems in hominid evolutionary history. He then cites Cachel (1994) who argues against the transferability of Machiavellian intelligence to nonsocial domains, to the creation of "natural history intelligence" where awareness of an attention to the nonsocial world occurs. However, this transfer seems likely to be seen in the history of religion and has been argued so by a number of students of the origin of religion, as in Hume's writings (Russell, 1988) or Muller's (1892) work.

Dunbar's theory is similar to the "Machiavellian Intelligence" theory of Byrne and Whiten (1988). Here, a feedback loop of increased fitness for clever means of dealing with increased social life resulting from complex social interactions of dense populations (and cooperative ones as the theory was modified by van Schaik et al., 2012) places early hominid evolution in the context of the evolution of the social life of birds. Van Schaik et al. (2012) also emphasize the culture component role, a theory proposed by Tobias (1971) and Bielicki (1964) as a means of accelerating evolution. This is best applied to vertebrates as social insects did not see any uniform increase in brain size as a result of increased social life, but should also apply within orders comparing nonsocial to social animals. Thus, the only outcome would be increased efficiency in completion or cooperation among conspecifics, but the increased fitness of the hominids is hard to define. Cultural complexity in this view means increased control or manipulation of conspecifics, thus the social brain evolves due to the rewards to fitness of manipulating other members of your species. We have no evidence of any complexity increase among hominids between *Homo rudolfensis* at 2 ma at about 700 cc of brain tissue and 500,000 BP with *Homo heidelbergensis* and no significant increase in population. Perhaps, this means manipulating females as Darwin (1874) saw the oppression of women as the model of domestication. But he also noted that the basis of this was sexual dimorphism, and other vertebrates show that one sex can dominate the other (e.g., *Lemur catta*) and still see no significant brain expansion. Hrdy (1981) produced a comprehensive assessment of gender across the animal kingdom.

Yet, ideas attempting to explain large brains in Cetaceans also address the social interaction necessary for survival and building networks among individuals, especially in social cognition of layer V spindle neurons, and we see a high ratio of glial cells to neurons as distinguishes human brains from monkeys and apes (Marino et al., 2007). Though dolphins form complex alliance relationships (Connor, 2007), Silva et al. (2005) found that dolphins engaged in a wide variety of sexual activities. Scott et al. (2005) argued that dolphins were highly charged sexually and aggression is often associated with sexual coercion, so perhaps hominid brain expansion is based on sexual exploitation or female objectification. (Figure 2.2 offers a comparison of dolphin and human brain evolution.)

Associated with the idea of group size are other theories that attempt to go beyond the size of the brain, and included here is the theory of the extended mind (Barnard, 2010) creating an interaction between "embodied cognition," language, and social interaction (Clark, 1999; Donald, 2002; Pfeifer and Bongard, 2007). There is in this concept an extension of the abilities of the brain into the social context and aspects of culture (e.g., tools), and all become part of the embodied cognitive field thereby extending the power of the brain into a more powerful mind. The main problem here is that we cannot test this theory. We cannot notice or verify any aspects of behavior in the fossil record that would support such claims. The brains of *Homo erectus* are bigger than early *Homo* (e.g., *H. habilis* and *H. georgicus*), while that of *Homo floresiensis* is much smaller. This theory could justify the idea of a

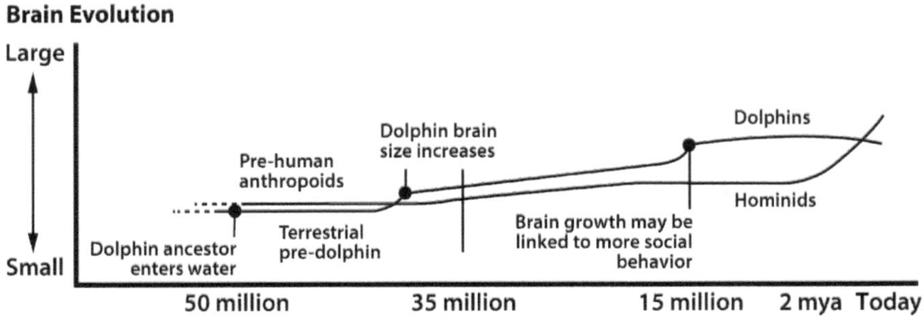

Figure 2.2 Dolphin and human brain evolution. ***Source*****: Created by the author. Drawn with materials from Marino (1996), Marino et al. (2007), and Foer (2015).**

smaller species being represented, as big brains and tool making do not seem necessarily linked, yet more evidence is necessary to support the contention that *H. floresiensis* is a separate species and not just a pathological example.

The discovery of a new find of early *Homo* at 2.8 ma (Shreeve, 2015a) or the idea that a new species, *Homo naledi*, has been identified only emphasizes this problem, as does the discovery of tools dating from 3.3 ma (Thompson, 2015). *H. naledi* (Berger et al., 2015) is likely to be a version of *H. habilis* or *H. rudolfensis* but the lack of a firm date is confusing. Cranial architecture seems to place it at the base of the australopithecine/*Homo* boundary (Schroeder et al., 2017). The fact that a number of individuals reflect australopithecine traits is not surprising, given the small number of early *Homo* remains heretofore reported. But, species and generic designation for hominids has been in flux for years and continues to be so (Schwartz and Tattersall, 2015). Yet, here also, Dunbar and Barnard's theories parallel that of Klein in calling for a mutation that produces an immediate and miraculous transformation of ability, and Chomsky's concept becomes again attractive.

Deacon (1997) argues against this on the neurophysiological basis, suggesting a long history of speech competence. One flaw in Dunbar's theory is the definition of relationships that quantify the rationale for group size (de Ruiter et al., 2011). Still, Dunbar's idea of these relationships (e.g., grooming and gossip) has parallels in the social behavior of dolphins (Marino et al., 2007). Unfortunately, Dunbar attributes significant motivation to animals (including early hominids) in an anthropomorphic and ethnocentric fashion; for example, he claims, "A group of this kind is an implicit social contract" (Dunbar and Shultz, 2007). This is referring to any of the higher vertebrates, but mainly primates where pair bonding and parenting take place in a social context. He goes on to assert that "members are necessarily obliged to trade off short-term losses in immediate benefits in the expectation of greater gains in the long term through cooperation" (Dunbar and Shultz, 2007). This tendency to couch animal choices in terms of the "rational man" theory is disturbing, yet he even seems to argue that primates are believers in "trickle-down economics" (Dunbar and Shultz, 2007). This idea does not even work with humans as the recent economic credit crisis of 2008–2011 demonstrated (Caldararo, 2009b). This is not to say that humans and other animals do not engage routinely in choices that can be assessed as "rational decision-making," which takes us back to what is rational or logical and whether this entails levels of maximizing (Etzioni, 1988).

A central problem here is what Antoine Wystrach (2013) has described as a major change in addressing the idea of insect cognition. New research has "revealed an alien complexity, one not driven by anthropomorphic considerations." Ants, unlike humans and most vertebrates, do not integrate stimuli into "a unified representation of the world, a so-called cognitive map. Instead they possess different and distinct modules dedicated to different navigational tasks." Ants do have memory and create systems of information to recognize territory and location (Wystrach, 2013). But as Barbara Webb (2012) finds, our ability to discover methods of cognition in insects is limited. Webb (2012) delineates the fact that neural systems evolved to transform behavior, not to represent it. Yet, the evolution of the neural cells, neural apparatus, the signaling molecules, and genes all show a consistent evolutionary sequence as Griffin (1976) recognized and has been more clearly defined in the past 40 years (Arendt et al., 2016).

Wystrach (2013) discusses insect navigation indicating both transformation (the animal responds to stimuli) and representation (it can locate itself in the world, a kind of map). As Bonner (1980) suggests and Webb (2012) describes with the desert ant, the animal modulates information to arrive at a successful return to home after chemical trails have evaporated, adjusting smell, light and other stimuli to a "home vector." A desert ant of the group *Cataglyphsis,* on returning home with food (using the "home vector"), can return to the place of the food. Does it have a "cognitive map" of the terrain in its memory or does it use an image? Bees can accomplish corrections to locations created by investigator misplacement of them. Webb (2012) uses the term "cognitive map" for this relocation information. The examples she describes of insect learning also parallel vertebrate learning achievements. However, using terms for human cognition, "conscious awareness," "understanding," "mind," or "intention" is confounding and creates inauthentic comparisons.

Dunbar and Shultz (2007) have also linked large brain size to pair-bonding as a relationship requiring considerable cognitive ability and rewarding the individual with higher survival value given the associated reduced risk in offspring survival. This is a rather ethnocentric view (along with their economic comments noted above), given that it implies monogamy is the only human-mating pattern and the data on pair bonding variation in other forms of consort are ignored (e.g., polygyny or polyandry). But even where Dunbar and Shultz (2007) compare birds to mammals, the chart they produce indicates very low pair bonding, which is contradicted by most recent data (Cezilly et al., 2000). Divorce (as in humans) and female aggression play a significant role in bird bonding, but the fact that most male birds lack a penis may be related to variations in mammals and birds (Herrera et al., 2013). It seems, however, if we look at insects, the idea of larger and more differentiated brains could be associated with eusociality. But when structures that appeared to be key to complex processing and larger differentiated insect brains (e.g., mushroom bodies) were studied across insect genera, it was found that these structures appeared before eusociality and were instead related to spatial learning and foraging, not sociality (Farris and Schulmeister, 2011). They are regarded as multimodal sensory integration centers (Wehner et al., 2007). This fits with the evolutionary history of hominids; collecting foods, avoiding predators, and an omnivorous diet could then be considered as preconditions for brain development and specialization.

Thus, we have a paradigm across genera leading to eusociality and behavioral complexity. However, differences among the social insects are considerable, for example, some scientists place an emphasis on not only reproductive specialization, but emphasize the idea of fixity in roles (Engel et al., 2016). This generalization can lead to distortion of the actual plasticity of role in many species, as where we know that, for example, honey bee division of labor can change distinctly between castes or be transitional where an individual can occupy two castes at once and demonstrate considerable phenotypic behavioral plasticity (Johnson and Frost, 2012). We must assume that this plasticity had an evolutionary history that is known in general outline, yet unclear in detail (Lin and Michener, 1972). Referring to insect brains creates problems in relation to vertebrates, as their "brains" in ganglia are not located all in the head region; some essential functions can continue after decapitation (Roeder, 1970), and the spread of the head "brain" is somewhat decentralized in the head-and-mouth region, rather like in other invertebrates such as cephalopods (Young, 1971).

We should also be careful of the seduction of cultural orientation to size. Some scientists have recently argued that "Natural selection favors an increase in the size of some organ if this increases lifetime reproductive success by improving the rate of survival or reproduction" (van Schaik et al., 2012). Gould (1977, 1996a) cautioned against adhering to such rigid progressivism and near-vitalism, given the fact that most organisms are small and, if such an assertion were true, all animals would be large and their organs also large. It ignores the reproductive risk of whole organisms as systems in complex environments.

A DEFINITION OF COMPLEX ANIMAL SOCIETY

Eusociality has a definition that applies mainly to insects. Batra (1968) coined the term regarding the cooperative behavior of Halictine bees, and this basic idea has been extended over time. Wilson's concept (1971) was constructed around the behavior of a set of social insects possessing:

1. Cooperation in caring for the young.
2. Reproductive division of labor with more or less sterile individuals working on behalf of individuals engaged in reproduction and the overlap of at least two generations of life stages capable of contributing to colony labor.

He considered eusocial to be the equivalent of "truly social" or "higher social." Since this introduction of the term, it has been extended to animal species of other orders, including coral-reef shrimp (Duffy, 1996), while mole rats, meerkats, and the dwarf mongoose are sometimes considered eusocial (Williams and Shattuck, 2015). Wilson's list seems too short to adequately define such a wide group of animals and differentiate it from simply colonies of single organisms. He refined this term in 2009 (Wilson and Holldobler, 2009), where "truly social" means

1. adult members divided into reproductive castes versus nonreproductive workers;
2. adults of two or more generations co-habiting the same nests; or
3. nonreproductive or less reproductive adults caring for young.

In this publication, they argue the equivalence of eusocial and superorganism. They also argue that the demographics and sociobiological traits of colony members are shaped not only by each individual's direct genetic fitness, but by the summed effects of all their performances on the fitness of their colony. They then refer to T.D. Seeley's (1999) statement that collective wisdom arises from poorly informed masses, whose interactions are shaped by natural selection among competing colony genotypes. This is perhaps a mistaken view, like analyzing the behavior of humans from a satellite, noticing them walking while looking at smartphones, and interpreting this as the sum total of knowledge. Yet, it is a perspective from quorum theory of swarm analysis where individual behavior is discounted, and certainly, from a phenomenological perspective, human behavior is also opaque. I discuss concepts of insects "thinking" in another portion of this book, but definitions of what is "thinking" are just as crucial as what is language.

It would seem that a more useful definition for eusociality as a term for complex social behavior would be

1. cooperation in food collection or production/sharing;
2. cooperation in nest construction;
3. joint brood care; and
4. colony permanence.

This definition can be used for vertebrate or invertebrate animal societies. Nevertheless, some degree of concern is required to make comparisons, as in the case of specific needs of animals, as in the case of surely vertebrate needs that attend human culture, as in having an endoskeleton and the need for clothing in certain environments versus that of an insect with in exoskeleton where such an addition would perhaps be maladaptive. (For reference in this discussion, figure 2.3 shows the antiquity of various adaptations to serve human needs.)

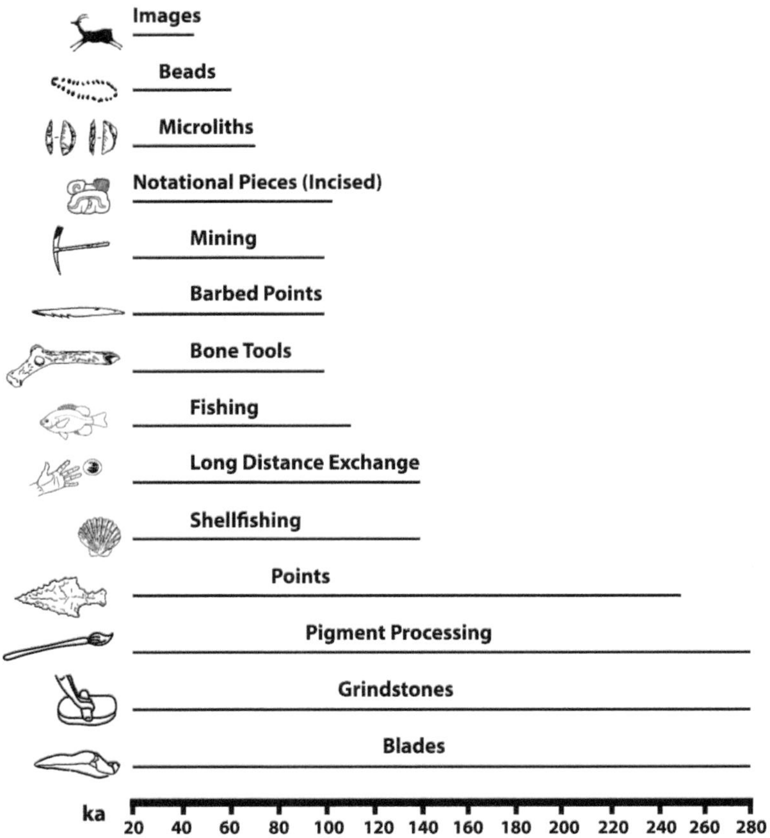

Figure 2.3 Human behaviors and their antiquity. ***Source*****: Created by the author. Redrawn from McBrearty and Brooks (2000). Used with permission.**

Burkart et al. (2009) used the term "hypersociality" to describe human society based on cooperation and cooperative breeding creating a shared intentionality leading to food sharing and group bonding. They identify cooperative breeding in other primates, but use other terms as in "prosocial" (behaviors that produce benefits for others) that could be synonyms for altruism as hypersociality could be for eusocial. We will use the term "eusocial" for animals whose social life is complex, as defined by Wilson and Holldobler (2005), and include food sharing, caching, and coordinated group behavior. They also use the term "primitively eusocial" to describe species whose worker castes have not become anatomically distinct. Human elites have often created separate workers to care for their young, usually nonbreeding females, but in recent decades, rich people have been hiring surrogates either to become impregnated with the sperm or to fertilize the ovum of an owner, give birth and then surrender the child. This comes close to a specialized reproductive regime and this process has advanced into various technological innovations to produce offspring for elites (Stone, 2010).

Duration of social structure was not significant to Wilson and Holldobler (2005), given most of the insect societies they included have periods of complexity collapse either as in some ants where environmental degradation requires a new location or as in bees where the bee population regenerates by swarming and a new queen. Human complex societies appear to undergo periodic loss of complexity as well although interpretation of this transition differs significantly (Caldararo, 2013; Chew, 2001; Tainter, 1988). It seems logical to include contemporary human society in the category of "eusocial." In a recent book, Wilson uses the term for human society (Wilson, 2012), so it seems to be an acceptable usage. But reference to insects and insect society is pertinent. Gowdy and Krall (2016) use the term "ultrasociality" from Campbell (1983), but just as "hypersocial" seems to indicate some abnormality, "ultrasocial" implies some idealized ultimate state of social life, which, I think, we have no justification to assert. Such a consideration might apply to termites where Lin and Michener (1972) attribute the "encouragement" of termite social evolution to their intestinal symbionts. Another term, "unicoloniality," includes "social immunity" at its core (Ugelvig and Cremer, 2012) and emphasizes the inclusion of all species members in a vast superorganism in which its success is not dependent solely on common defense and exploitation of resources but also on organized hygiene and care of sick colony members. They also argue that eusociality results in rigid social roles and a lack of individuality, but this is contradicted in prairie dogs (Slobodchikoff et al., 2009) and social spiders (Laskowski and Pruitt, 2014). Social spiders' behavior (Brach, 1977) conforms to our definition of complex animal society and fits the original intent discussed by Batra (1968).

For humans, we want to distinguish ideas of modernity from eusociality as defined here and in some cases for the behavior of anatomically modern humans (AMH), as some behaviors are used as special markers, like art—yet *what* is art is quite subjective and culturally bound. Whether behavior-creating art is modern or not is questionable and cannot be associated with only the peoples of the Upper Paleolithic, as the examples of the Venus of Tan-Tan (200,000–500,000 BCE) (Bednarik, 2003) and the sculpture at Berekhat Ram (200,000–700,000 BCE) (d'Errico and Nowell, 2000) demonstrate. Even at the earliest potential dating of these, we are not dealing with AMH, yet we also have widely scattered examples of "modernity," for example, in Asia by *H. erectus* in engravings in Java (Joordens et al., 2015). Also, the problem of taphonomy is central to our interpretation of human ability in the past, as what is preserved and what we can recognize of human consciousness are two main factors. Another consideration is that small populations produce fewer objects that can be preserved.

In the Hymenoptera, the mushroom bodies (see figure 2.4) are found in highest number and the packing density of microglomeruli is greatest of any insect order (Groh and Rossler, 2011), though there is a wide variety of mushroom bodies in various invertebrates and some disagreement as to their relation to behavior across this large group (Strasfeld et al., 1998). They hypothesize that this combination may increase or enhance computation capacity. Hofman (2014) makes a similar argument for humans. One might argue that the hominid neocortex, like the Hymenoptera structures, has become elaborated as a consequence of computing special relationships. This allows for different behavioral specializations as Farris and Schulmeister (2010) describe, comparing the large mushroom bodies of some beetles with those of wasps, bees, and ants.

Haug discovered that "The comparison with man in columns of brain section from the cortex shows that the mammals with large brains have mainly large cells of the pyramidal type in their cortices, while in the size classes below the projection area of 100 pm^2, man has a lot of small cells which are probably mostly of granular types. The density of neurons in man is distinctly higher than in mammals with large brains. The elephant, with 6.8 ($\times$ 1,000/mm^3), has the lowest Ne-de measured in our investigation" (Haug, 1987). But Groh and Rossler found that "Synapsin-immunolabeled whole-mount brains reveal that during the natural transition from nursing to foraging, a significant volume increase in the calycal subdivisions is accompanied by a decreased packing density of boutons from input projection neurons" (2011). Similarly, we find in mammals differences in neuronal cell types, numbers, location, and density and that these vary with the size of the animal, but they are associated with behavioral complexity, as we see in the differences in neuronal packing and projections in bees and, in this case, differences that

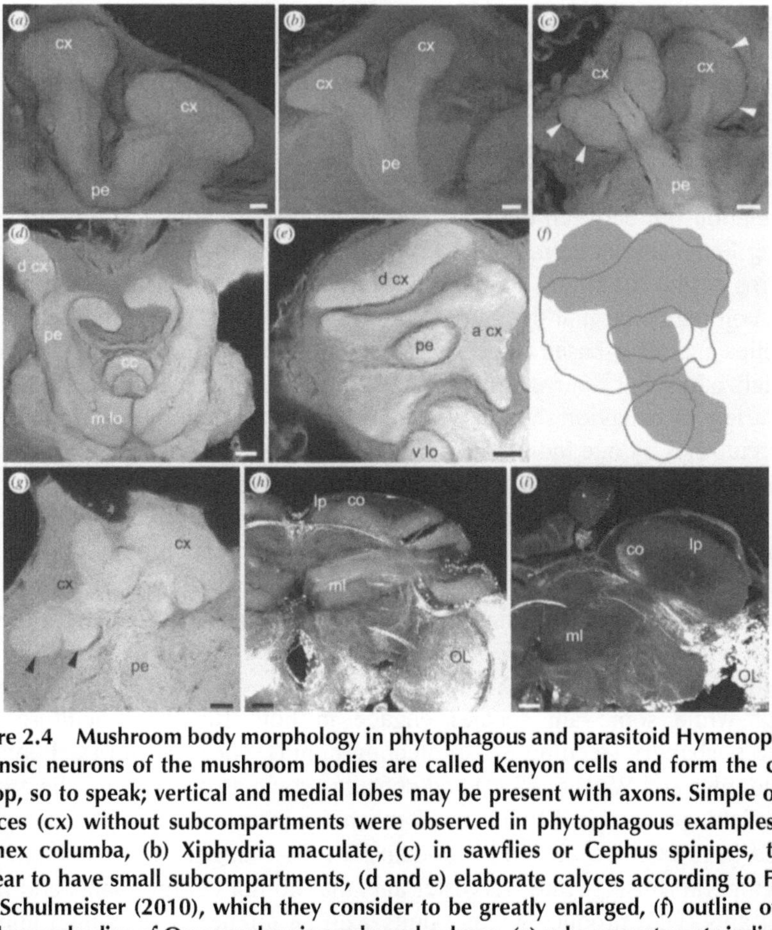

Figure 2.4 Mushroom body morphology in phytophagous and parasitoid Hymenoptera. Intrinsic neurons of the mushroom bodies are called Kenyon cells and form the calyx or top, so to speak; vertical and medial lobes may be present with axons. Simple ovoid calyces (cx) without subcompartments were observed in phytophagous examples: (a) Tremex columba, (b) Xiphydria maculate, (c) in sawflies or Cephus spinipes, these appear to have small subcompartments, (d and e) elaborate calyces according to Farris and Schulmeister (2010), which they consider to be greatly enlarged, (f) outline of the mushroom bodies of Orussus showing enlarged calyces, (g) subcompartments indicated by arrows of elaborate calyces of Stephanus serrator, a basal parasitoid suggesting subdivisions by olfactory and visual inputs, (h) optic lobe dextran fills show visual input to collar of calyces of the parasitoid Gasteruption sp., (i) mushroom bodies of Leucospis sp. possess single flask-shaped calyx that receives optic nerve projection neurons to a small ventral collar (co). pe = pedunculus are generally parallel axons of globuli cells that extend to the front of the brain where axons then branch to provide a vertical and a medial lobe. Interpretations of mushroom bodies, their components and variations differ and they seem to function differently in the invertebrates in which they have evolved; intraspecific differences have been correlated with social hierarchy, foraging, and other behavioral experience (Farris and Schulmeister, 2010). The complex of structures has been likened to the cerebrum and neocortex of vertebrates, and in both groups, one would expect their variations and complexity to be a result of a variety of evolutionary factors and selection scenarios. *Source*: Farris and Schulmeister (2011). Used with permission.

are significant generationally. We know that in humans, there are significant changes in brain size and neuronal survival with age (Haug, 1987) and that this is generally associated with a degeneration of behavioral complexity, or dementia (Burgmans et al., 2009).

One additional concern might be made regarding the size of different components of the insect brain and social life. Riveros et al. (2012) examined the relationship between the brain size of fungus-growing ants and sociality testing Farris and Schulmeister's (2011) thesis that there was a relationship. Riveros et al. (2012) make a number of assumptions regarding sociality, the most colored being that there are, in their opinion, two types: individualized societies and class-based societies. Essentially, this focuses on the degree of specialization, but their treatment of the varieties of social insects obscures the variety of behavior, though we often see the production (performance) of different roles by one individual during its lifetime: in some social bees, care givers become foragers, and in termites there is a significant plasticity in role. Their comparison with human societies is also concerning as their generalizations from the literature (Naroll, 1956) would indicate a more determined relationship (limitations) in human social evolution than is supported by ethnological research, for example, dispersed Mayan settlements from monumental sites and undiversified living quarters, and unstratified cities of the ancient Indus civilizations (Sabloff, 1996).

But while some ant species engage in both fungus agriculture and active husbandry (feeding and protecting aphids, coccids, and treehoppers) (Holldobler and Wilson, 1990) and consume some as do humans, the evolution of this process is difficult to establish for insects as the antecedents are buried in pre-Cretaceous eons. Godzinska (2016) suggests that evidence of contemporary insects like Harvester ants that do not culture plants and honeypot ants (who produce storage units) can be aligned with human hunter-gatherers who also produced storage containers, and there is evidence dating prior to 60,000 BP (Texier et al., 2010).[1]

We have investigated many of the main explanations for the evolution of mind and consciousness in *Homo sapiens* and in other eusocial animals. It seems that the origins of "mind" and consciousness are not definitely linked with any specific physiological conditions (e.g., large brains). But social complexity may be a consequence of certain combinations of selective pressures like predation, food distribution, and disease with which increased brain size and behavioral plasticity is a positive factor. Our next task should be to examine the nature of social life, its benefits, and consequences.

NOTE

1. I cite other examples in my 2014 text (see References).

Chapter Three

Group Size, Territory, and Disease

Abstaining from wrong judgment would make life impossible.

(Nietzsche, 1889)

Why does the human brain have to be so big? As Richard Alexander writes, "It causes us to wonder what our brains were designed to accomplish, and to suppose that there are no challenges in everyday life that are sufficient to explain them" (1989). (For reference, figure 3.1 shows the human brain in comparison with those of other vertebrate species.) The central question to ask is, "What selective advantage is gained by constraints of brain size to group size?" Holloway (1996) found that the differences between Neanderthal brains and modern human (anatomically modern humans [AMH] or *Homo sapiens sapiens*) brains were very small, except that Neanderthal brains were slightly larger. In studying Neanderthal sites and comparing them to modern human sites, it was found that Neanderthals tended to utilize territory differently than AMH (*Homo sapiens sapiens*) and this had significant effects on survival, and also that it could be linked with perceptual differences as well (Pearce et al., 2013). In my estimation, the differences were less significant and more likely to be the result of cultural differences or even seasonal variations of use as Coles and Higgs (1969) suggested from similar evidence. Though as Holloway (2004) notes, generic and specific names are not written in stone and have been changed over time under reanalysis.

Dunbar and Shultz (2007) argue that behavioral ecology of primates is related to the omnivorous diet and widely dispersed behavior of hominids over less diverse diets of some primates and localized behavior. That this would result in increased selection for increased brain size would, however, reach a limit as described by Aiello and Wheeler (1995), given the expensive

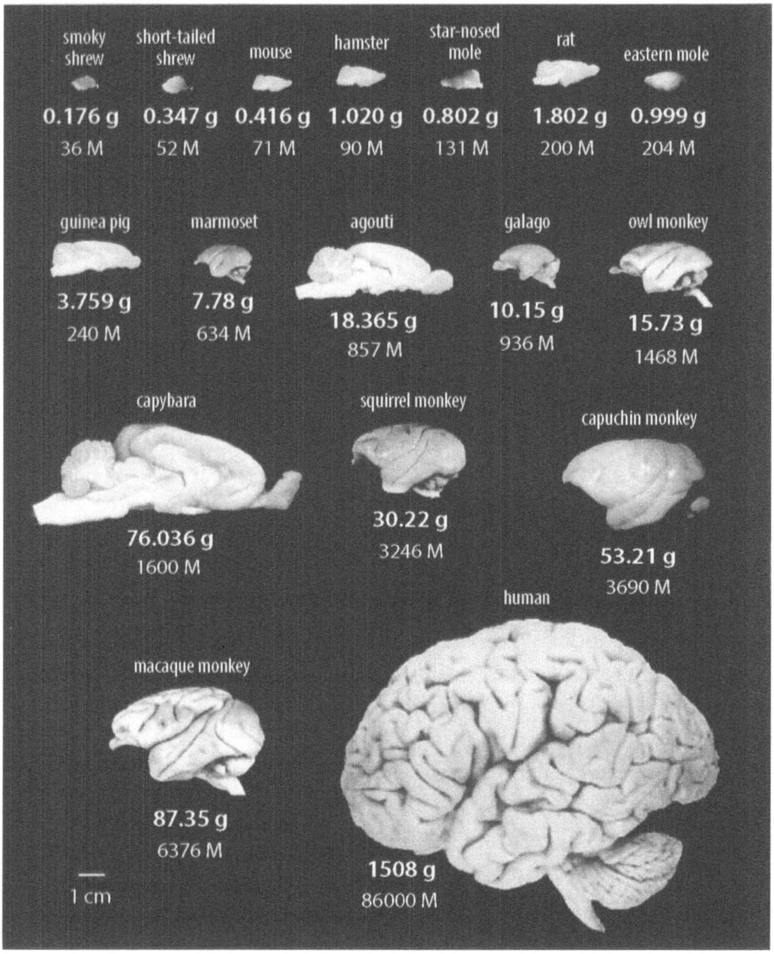

Figure 3.1 Brain mass and number of neurons for various species.
***Note*: Brains are arranged left to right and top to bottom in order of increasing number of neurons. All images are to scale. *Source*: Herculano-Houzel (2009).**

tissue of brains. This size limitation factor would have to be balanced to some selective advantage, perhaps as seen in bats, where Megachiroptera with its larger body size and larger brain can be compared with smaller-brained Microchiroptera evolving in competition with their prey (e.g., insects) and predators (given the benefits of sociality, roost election, and site modification) (Baron et al., 1996). But the increase in brain size in hominids from *Homo habilis* to *Homo sapiens* cannot be explained by any objective analysis of selective factors. There was no significant population increase from 2 ma to 50,000 BP that parallels the increase in brain size, and no dramatic specialization in behavior until 10,000 BP, when we do see beginnings of population expansion associated with domestication technologies. In fact, Barton (1996, 2012) focuses on the relationship between the expansion of the cerebellum and types of ecological strategies (Barton et al., 2011). This is argued to have had a significant effect on Neanderthal survival at about 30,000–40,000 BP or just before, as opposed to *Homo sapiens sapiens*. Yet the cultural capacity of these two populations cannot distinguish the nature of humanness or any significant appearance of technology that might indicate brain reorganization (D'Errico, 2003). Humans have displaced each other from territories for the entire historic period and this continues today, yet no one would argue that Viking Norse, Anglos, and Saxons were humans and that those they displaced, the Scots or the Irish, were not.

Dunbar et al. (1999) also note that the human infant is born with only a portion of its brain developed, (about 24 percent according to Pena-Melian, 2000, or perhaps more; see DeSilva and Lenik, 2008), and this certainly creates a context where the social environment is "imprinted" on every vocalization and facilitated (rewarded) gesture and context (see also Deacon, 1997). We might argue that this system is where language began to hold attention and represent time and space via that social imprinting and context, though more recent research has shown that while it is true that human infants are born with significantly undeveloped brains compared with other primates, chimpanzee infants show a significantly slower rate of development than human infants (Sakai et al., 2012). Compared with the rat pup brain that reaches 90 percent of its adult weight by postnatal day 20 that corresponds to 2 to 3 years in humans, the human brain undergoes substantial variations in cortical region, thickening, thinning, and synaptic pruning to a white-matter peak at 37 years and gray matter stabilizing at about 50 years (Semple et al., 2013). This undermines earlier ideas of human neonates being born with all the neurons they will have as adults (Sherwood et al., 2006), given also that we find reports of continued adult neurogenesis as well (Paredes et al., 2016). But Paredes and associates also found that the increase in brain size and topographic complexity may severely limit the contribution of new neurons born in or close to the ventricular wall. Sherwood et al. (2006) also found that the glia–neuron ratio does not differ significantly in anthropoids (including humans), but they report a greater metabolic consumption of human neocortical neurons relating to expansive dendritic arbors and long-range projecting axons.

It is only when we reach the Neolithic, where we find changes in technology affecting ecological exploitation in domestication and structures, that a clear demarcation appears. At this point, we have an increasing density of population and degradation of the immediate environment, mainly increased toxicity by the accumulation of wastes from humans and animals with water pollution perhaps being the most significant result. Herd diseases from domesticated animals and toxicity of spoiled food and parasites would add to the new combinations of pathogens and selective pressures of sedentism (McNeill, 1976). Disease would then attack both smart and dumb alike. Add to this the element of conflict, both internal clan and village and then city versus city warfare, and we see that selection for ability would be seriously affected. Drop a bomb on Dresden or Hiroshima and you kill all across the IQ spectrum. What began to change then, 10,000 years ago, was the selection for general intelligence against rising intelligence, as individual increases (genius) would be too expensive a gamble and a waste of resources given the probability of death due to chance events and mortality like disease and violence (Caldararo, 2012b).

Woodley et al. (2014) argue that the development of intelligence in monitoring signs of disease and infection among other hominids as well as other animals would have survival value, as I have also developed more broadly (Caldararo, 2012b) for many animal species, especially those in complex insect societies (Caldararo, 2015a). Also, disease has been a significant force in shaping human social interaction, not just in avoiding disease and developing behaviors to accomplish this, but also in building social mechanisms to organize individuals to learn techniques to reduce septic conditions and infectious outcomes. We see this also in complex animal societies (Heinze and Bartosz, 2010).

Some authors cite work on IQ and brain size (McDaniel, 2005). The McDaniel study was a metastudy; its sample was cherry-picked by the author, and the female sample did not produce parallel results with the male. Also, there was no correction for brain/body ratio, the methods used to consider intelligence were unclear across the studies cited, and it was uncertain how the tests were conducted; since they were often of different types, it is unlikely they measured the same abilities and are comparable. A similar problem plagues the Flashman (1997) study which found only "modest" association between brain size and IQ. Schoenemann et al. (2000) found that brain size did not predict general cognitive ability within families and are critical of other studies. Some studies find socioeconomic status has a significant effect in IQ heritability (Turkheimer et al., 2003). Other studies of IQ are marred not only by the history of fraud that Gould (1996b) has listed but also by association studies that lack explanation, as in that concerning formin binding protein 1-like (FNBP1L) (Benyamin et al., 2014). While association is not causation, FNBP1L is involved in antibacterial autophagy (Huett et al., 2009), and its appearance in some people with generational ability does not tell us if the gene for the protein is involved. Yet the search for proof of the inheritance of intelligence goes on, as Gould (1996b) and Montagu (1964) have shown; it has been often plagued with racist, classist, and sexist thinking.

It is likely that ability, or the "g factor" ("g" usually refers to general intelligence [Deary, 2010], though there is disagreement on its nature and measurement; see Brody, 2008), increased during the Pleistocene, but that after the sedentism of the Neolithic, epidemic and endemic disease began to level selection pressure on brain size, due perhaps to an upward limit on maintaining such expensive tissue (Aiello and Peter Wheeler, 1995); war would be another leveling effect to bring about selection for general intelligence, as the brightest and healthiest would be exposed to early death before or at greatest reproductive potential.

Both Gould (1996) and Montagu (1964) address the inherent vice of intelligence tests and such assessments in general. Subjectivity seems to dominate the field, as Gould and Montagu have demonstrated. The big brain became a central focus for paleoanthropologists as it had been for biologists and anatomists for centuries. Krantz (1961) and Tobias (1971) claimed that the size of the brain in children at the end of the first year of life (approximately 750 cc, see also Holloway, 1996) should be the meridian when a brain size at this point indicates the arrival of symbolic behavior. This is entirely based on the assumption that the human child today creates symbolic language at this time and with this amount of brain. Yet humans do produce speech with smaller brains and the language of children at 12 months can hardly be said to represent a fully achieved human consciousness. Some studies of language acquisition fail to distinguish between ages of children studied and use terms like "children," "young children," and "very young children," and compare these unclarified results with single primates or at best few (Yang, 2013). The child gains brain size and neuronal numbers up to about 90 percent of adult size by two to three years of age (Semple et al., 2013). Still, the capacity of a child at this age can hardly be said to be fully human. By ages six to eight, this level of performance usually does arrive, given proper exposure to human society and nutrition.

Nevertheless, for Tobias (1971) and Krantz (1961), there had to be some explanation for so much brain tissue added to the hominid brain over so long a period of time but with so little complex behavior leaving traces. How could this expensive tissue be justified? Essentially, this is an argument based on a materialistic foundation. No material change in behavioral residues must mean the animal is not capable of complex cognitive life. Here Griffin's (1976) complaint about behaviorists can come to play; if we extend the internal life of animals beyond the minimal usually accorded them and widen it to that of early hominids, the lack of residues of complex behavior is less of a problem. Dolphins produce few tools as do cephalods, yet their capacity for a rich and complex internal life of thought seems a likely consequence of the evolution of a large brain. Here a joining of the social brain and environmental brain hypotheses produces a potential solution.

However, social complexity and domestication are related in human history and the process of domestication of animals was related also with the evolution of sedentary life producing the opportunity for cross-species infection (Fabrega, 1974, 1997). Other animals who are social who possess food production and dense populations have evolved or developed social behaviors to reduce septic conditions and infection as well and include bees and ants where pathogens interact in living contexts with dense populations of hosts (Caldararo, 2012b). In bees, we find this takes the form of cleaning the hive (Rothenbuhler, 1964), and there are eusocial bees without this behavior.

What has not been clear is that these behaviors and diseases were all part of the induction of human eusociality after the last glacial period. That domestication brought humans into close proximity with animals as never before, and the exchange of pathogens must have been enormous. We can imagine this effect by reference to the horrific results of the arrival of Europeans and their diseases on Native Americans (Merbs, 1992). The fact that ant brains evolved concurrently with a parasitoid mode of life and see the appearance of large bodies associated with neurons (Farris and Schulmeister, 2011) is parallel to the fact that humans have a larger neocortex than other primates, and the two phenomena may explain the dramatic appearance of complex behavior after 10,000 BP in humans as no other factor has provided (Faris, 2013; Strausfeld et al., 1989) but the fact that mushroom bodies seem to have appeared in insects long before social complexity (Farris, 2013) is parallel to the fact that humans have larger neurons in some cases than other primates and larger brains long before the appearance of social complexity. But some other factor must relate these similar trajectories, and in humans, this parallels the insect situation in the late and dramatic appearance of complex behavior after 10,000 BP in humans.

Arguments that territorial defense or kin selection can account for the appearance of eusociality seem inadequate given the remarkable variations seen in the insect world (Grinsted et al., 2012; Lin and Michener, 1972). Gowdy and Krall (2016) emphasize economic principles and cite Diamond (1997, 2005), whose environmental determinism closely parallels their own economic determinism. They have, as they mention in the commentary to their paper, departed from the usual narrow economist perspective in their comparative approach. Gowdy and Krall (2016) refer to economic configurations of diverse species but fail to produce comparisons that would identify these, and they argue these configurations "around agriculture" look the same and that social behavior in humans and insects "developed similarly as groups once they began the transition to agriculture." Surely this is unknown. We do have few fossil remains of eusocial evidence for insects (Engel et al., 2016), but an ethogram listing the appearance of these behaviors and a time frame for their integration into social structures or even genetically coordinated complexes are missing.

As for human origins of eusociality, the picture is not as clear as Gowdy and Krall (2016) argue, even if we assume that the process was underway at the end of the last glacial period, as proponents of niche construction theory (NCT) suggest (Laland and O'Brien, 2010). If we assume some 9,000 years of development until we find evidence of early sedentism, which some place earlier at 15,000 BP with sedentism developing without agriculture (Dow and Reed, 2015), though variations to this transition are rife (Bar-Yosef and Belfer-Cohen, 1989), the picture we have of the evolution of the use of space and origins of urbanism is still unclear (Finlayson et al., 2011; Hodder, 1978). Their idea of uniformity of the economic configuration or similarity of groups once begun on the transition to agriculture lacks sufficient data to make the argument. We are confronted by events at the end of the Bronze Age in Europe with an "end point" vision, with an interpretation of present as past. The same lack of information regarding institutions applies to criticism of Gowdy and Krall (2016) by Houdek, Novakova, and Stastny (2016). Houdek et al. (2016) argue that institutions in preagricultural societies would form the foundation for states and agriculture but debate over what has been considerable as an institution in recent years (Hodgson, 2006), and projecting this uncertainty into prehistory does not appear to clarify the origins of agriculture or the state, as state formation is complex and dependent on a number of cultural and economic factors (Caldararo, 2016a). However, Houdek et al. (2016) also attribute the late development of agriculture in Northern Europe to its greater wealth without reference to the fact that both the Industrial Revolution and the wealth they so admire were based on the slavery of Africans and the colonial oppression and expropriation of non-European native lands (Caldararo, 2013; Chang, 2008; Worster, 1988).

Gimbutas (1989) has argued that there is as much evidence of different forms of social life in the earlier period that can outline a more matrifocal (women-centered) structure, and the organization of Spartan life with government nurseries and communal life as a focal point of values is a stark contrast to that envisioned by Gowdy and Krall (2016). That Sparta had greater power and endurance than Athens is obvious from its victory in the Peloponnesian War. But, as Xenophon who lived in Sparta notes, while Sparta defeated Athens, victory defeated Sparta, since it had no institutions or patterns to guide the creation or management of the empire.

In all, Gowdy and Krall (2016) emphasize hierarchy and specialization in waves of division of labor that regiment early human society. However, a case can easily be made for heterarchy (Caldararo, 2013, 2014; Crumley, 1995) and primitive democracy in earliest Mesopotamia and Egypt (Jacobsen, 1943; Wilson, 1951).

The economic principles of Gowdy and Krall (2016) as they apply to insects may have been a driving impulse to eusociality in termites, the group which appears early as mastering eusociality, but not in all insects; eusociality developed differently in ants, for example. Ants appear to have been predators on termites and not eusocial as early (Engel et al., 2016). Lin and Michener (1972) suggest this pressure of predation is also a force toward eusociality as is the acceptance of "joiners" to increase genetic variation. There is a greater degree of plasticity in termite behavior than in ants generally (Roisin and Korb, 2011). Gowdy and Krall (2016) generalize when referring to insects, and so while "fungus-growing" termites usually refer to the Macrotermitinae, a subfamily of the Termitidae, with 14 genera and 350 species, it is often difficult to pin down references to specific species or genera possessing the traits being discussed. They also present a comparison of insect societies that have evolved over millions of years since the Cretaceous with the recent human society. Their pessimistic view of human future is also colored by their view of insect societies as being dysfunctional. Yet insect eusociality is well adapted and has survived horrific changes (including the end of the Cretaceous period's asteroid hit), and unless humans destroy the planet, these societies will likely continue as before human appearance. Gowdy and Krall (2016) also depreciate the ability of insect individuals to influence social direction or adaptation and equate the human loss of individuality to a parallel condition. However, insect members of eusocial groups have been found to teach each other, and such influences would produce the means for positive feedback to group behavior, even if we lack the means to track and contextualize such behavior and effects (Franklin et al., 2012; Franks et al., 2015).

Some critics of Gowdy and Krall (2016), like Levine (2016), reject comparisons of human and insect eusociality due to their own belief that the vast variation and plasticity of human behavior makes such comparisons a "fun house" mirror. Yet, if we consider the scale and the differences in physiology, we might find more than useful parallels. Gellner (1989), for example, describes how human societies present a dichotomy of social behavior, as humans are remarkably well disciplined and restrained within societies, but that great differences in behavior and social organization are found between societies and cultures. Hutchinson (1967) discusses the long history of economists and philosophers, whose works influenced government policies in Europe, America, and the colonies, debating how far wages could be reduced and still have an effective workforce, or increased and provided incentives. Muri (2004) and Thompson (1980) analyze the process of time management and regimes of time in work and life in Europe after the Protestant Reformation. All these have, including reading and the printing press (Goody, 1987), substantially led to uniformity and regimentation.

The overemphasis we see in some of Gowdy and Krall's works (2016) relates to the reduction of variation to a generalization as in the rigidity of castes in ants and dependence on close genetic relatedness, which has been disputed (Wilson and Norwak, 2014). The lack of castes in social cephalods and spiders is also an exception, yet they do not produce food. The reliance of Gowdy and Krall (2016) on "economic principles" is, perhaps, a result where biology is ideologically defined rather than empirically discovered. It is an aspect of progressivism and its forerunner vitalism to impute design to process. Wendt (1963) provides an excellent review of the development of theory in science. While dialectical arguments are often effective heuristic devices, and can be seen to "explain" historical events, they often assume a determined process, as in the case of Okasha (2006), on which they rely. Ascribing such movement to organisms would result in a picture of evolution where no species would ever become extinct, all genes would become equally "selfish," and economic behavior would achieve perfect rationality, even in humans! Rather, it seems the evidence of human history supports ideas of cyclical, rise-and-fall scenarios with humans learning very little from their history, economic or otherwise, but technology seems to speed the process (Caldararo, 2013).

Gowdy and Krall (2016) make a number of assumptions, one of them being that no insect species or society has ever reversed eusociality. Since the evidence for the origins of eusociality in insects is fragmentary and limited at best, this is a rather incautious assertion. For humans, it is obvious that human societies have periodically become less complex, as from the Roman period to the European feudal society. Their argument that human social evolution to eusociality has been detrimental to human health and happiness, as well, is also based on certain contemporary views of the "good" in social life (Caldararo, 2004a) and is of no significance here, as it is nearly impossible to find agreement concerning what is "good," for example, long life, children, gold stores, land, and so on. Also, we cannot know the immediate effects of incipient agriculture 10,000 or 12,000 years ago. It may have been remarkably "happy," but I will discuss the health relation with diet and the archaeological data later. Here I will only note that legends regarding agriculturalists and pastoralists or hunter-gatherers are seldom positive, as in the Biblical conflict between Cain and Abel. Forcing a rural population into cities is not uncommon as was seen with England and the Enclosure acts described by Polanyi (1944) in the eighteenth and nineteenth centuries, but enslaving hunter-gatherers as urban workers possibly dates to Sumer (Kramer, 1961). Thus the process of domesticating plants and animals leads to the domestication of humans. The question then is, "Could the agricultural revolution have taken place without forced urbanization, or could the industrial revolution have developed without colonialism or mass slavery of Africans?" We might consider that humanity is already organized into castes and classes. Baptist (2012) demonstrates how the wealth on which the Industrial Revolution was based came largely from slavery and colonial exploitation. Human complex societies have not always been based on slavery or hierarchically organized, and heterarchy has been perhaps a longer-utilized system, which I will discuss in another place in this book.

Other central points to their discussion are more curious in how they relate in contradiction to their theory, for example, the emphasis that agriculture was the transitional element to ultrasociality (eusociality), while they acknowledge that human sedentism occurred first, as among the Natufians or at other locations like Catal Huyuk, where people lived in close conditions and kept themselves clean (as ants or bees do in their colonies), and yet evidence of hierarchy is few (Hodder, 2011). They argue a uniformity for human society—one that would group the people of the ancient Indus civilization with that of Sumer, early China, and the Inca. While evidence of contact with Native Americans shows an integration of nomadism and planting corn with a return to harvest, this variety hardly fits their model (Hurt, 1987). Compare this to Julius Caesar's report on ritual warfare among the tribes of Gaul, where the constant yearly conflict could be seen as an adaptation to population control. Nevertheless, their reliance on Campbell's (1974) idea of "downward causation" is not the only interpretation for state formation and renovation. In Japan and China, rebellions often came from peripheral groups, as in the Han rebellion by minor official Liu Bang or the later rebellions (e.g., Yellow Turban Rebellion), which arose from the social base. Such renovations of China's social life (Caldararo, 2013) or in the rise of the Japanese state (Caldararo, 2016a) were common. Also, their requirements in terms of hierarchy and states mandate certain elements, like private property, yet Polybius (200–118 BCE) tells us that while Spartans were given equal amounts of property to farm, they lacked money. Land holdings could not be enlarged. Was Sparta then not an "ultrasocial" society? Also, much of the land holding data about Sumeria seem to contradict this idea and point to a variety of ownership as well as redistribution (Adams, 1966). Yet, there is evidence of rising inequality in the Sumerian cities at their unification and suggestions that Sargon achieved control by appealing to the dispossessed and those feeling oppressed or disenfranchised (Lewis, 1980).

Nevertheless, many writers seem to view their concept of complexity in insect societies (whether eusocial, hypersocial, or ultrasocial) as a static endpoint development. However, not only do colonies like termites adjust population density to food resources, but also when migrating to new habitats it appears that colonies of ants can lose symbionts that are responsible for a variety of social effects, host reproduction, fecundity, and male-killing (Reuter et al., 2005). Ant species are constantly evolving and modifying their social habits as conditions require, creating more adaptive social integration in their environments. This is, perhaps, what we see with human society, though reference to Polybius's theory of the limits of human government, or anacyclosis, seems to argue against this idea.

Chapter Four

Performing as Human or as a Social Being

In his lecture "Explanations and Applications," Freud observed, "We realized that the difficulty of childhood lies in the fact that in a short span of time a child has to appropriate the results of a cultural evolution which stretches over thousands of years" (Freud, 1966). Perhaps our focus on IQ has been misplaced: we recognize that it is related to nutrition and that, as in Bogin (2001), deficiencies in growth can rebound under certain circumstances. On the other hand, we also know that such assessments have been hampered by cultural bias in their construction and administration (Gould, 1996b). Today many researchers who focus on intelligence realize that it is "always displayed in a cultural context" (Sternberg and Grigorenko, 2004). Understanding that context and defining means of assessing cognitive skills within that context is the dilemma of IQ test fabricators. However, it also weighs heavily on any evolutionary perspective on the evolution of what it means to be human and how to assess that nature. Even one of the most used and considered "culture-neutral" tests, the Raven's Progressive Matrices test, has been shown to be sensitive to "stereotype threat" for both children of low Socio-Economic Status and girls (Desert et al., 2009).

This is an essential parameter for understanding the nature of ability. What is the context of becoming human? In his examination of the past hundred years of animal cognitive research, Macphail (1982) dissects the environment of experiments of their design and the assumptions of the experimenters. An important question brought up by Macphail is the problem of context in the experimental setting. This focuses on the natural orientation of the animal, not just regarding their sensory apparatus or physical organization of the body, but the maturity of the social behavior of the animal as well.

Testing for ability can be significantly distorted if these considerations are not applied. In this view, since Deacon and others regard language as an essential element in human behavior, the social context is doubly significant, as without language they regard hominids as not fully human. Macphail (1982) considers the idea that human intelligence is mediated entirely by language and thus might make comparison with other animals impossible, but investigates the possibility raised by some psychologists of using humans born without access to learning language as a surrogate, yet this has not been a fruitful area of research for a number of ethical problems. Therefore, how do we adjust for the problem of language, we cannot raise a human without it and test for ability, yet we do essentially the same thing with most animals, they are either raised in asocial nurseries, or kept in socially deprived contexts in comparison with natural settings. Here the problem is important when we compare highly social animals like *Cynomys* and *Castor canadensis* or *Suricata*. If the social context is essential for becoming human, then the same must be an element in becoming a beaver or meerkat.

BRAIN SIZE EVOLUTION AND SOCIAL BRAIN

There is a divergence of opinion regarding what change in the human brain could account for the significant increase in behavioral complexity of the past 7,000 to 10,000 years (figure 4.1 illustrates changes in cranial capacity over preceding millennia). Holloway et al. (2004) placed emphasis on a "reorganization" or "rewiring" of the brain. This makes sense, as simply the increase in brain size to body ratio is not associated with any significant change in human behavioral complexity after 150,000 BP until about 10,000 BP. This parallels ideas of connectivity in convolutions (Hofman, 2014). Here, then, we assume that what has occurred has been inherited by every human fetus as every normal human develops into an adult who can learn language, manipulate culture and innovate. This is certainly a remarkable evolutionary achievement, but it falls down when we realize that not every human inherits the same abilities and temperaments. The resulting attempts to salvage the theory are reduced to a statistical argument of bags of genes shuffling endlessly into combinations that still result in a general species "ability" that covers competency in language and most "skills," and yet is "iced" with a variety of more complex "abilities" like math and music. Specific genes that are found to have significant roles in language, like FOXp2, have been traced to about 0.5 mya, too late for Acheulean technology and too early for any culturally significant transition of the Middle Paleolithic, except perhaps Levollois technology (Schreiweis et al., 2014).

In the realm of physiology, the idea that more convolutions of the brain meant higher cognitive processes was commonly held. It was also believed

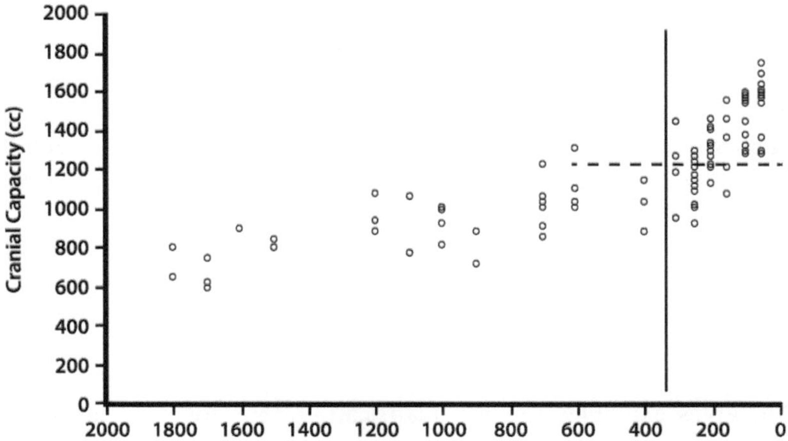

Figure 4.1 Cranial capacity (in cc) as a function of time for 94 specimens of *Pleistocene Homo*. *Note*: Vertical line indicates *Homo sapiens sapiens*. The dates of specimens between 1,800 Ka and 300 Ka are rounded to the nearest 100 Ka, and the nearest 50 Ka for specimens between 300 Ka and 50 Ka. *Source*: Created by the author. Redrawn from Lee and Wolpoff (2003).

that the cerebral cortex was the central region of the brain that controlled complex cognitive processes in mammals (Geschwind and Rakic, 2013; Kaas, 2013). Variations of convolutions were recognized across genera, but some nineteenth-century comparative anatomists were not convinced that gross numbers or shapes of convolutions were related to cognitive abilities or that there was a direct longitudinal series in evolution, as there were well-convoluted brains in many orders, while others with "presumably higher organization" have brains only "feebly convoluted" (Turner, 1890). Orr (1966) simply noted that convolutions were the result of increased nervous tissue "from lower to higher forms." Early researchers also discovered problems of fixation or preservation of brains that distorted surface features and size, and age was a factor (Connolly, 1950).

Still, research on the brain has focused on the cerebral cortex as the "biological substrate" of human higher cognition (Geschwind and Rakic, 2013). This was partially due to findings that brain-to-body-mass ratio does not always correlate closely with behavioral complexity and intelligence (Striedter, 2005), although as Macphail (1982) notes, many of the experiments assessing ability were tainted by assumptions over the nature of intelligence, how to design them, and what animals to use as models for different phyla. Yet, this selectivity of what animals to use or how to interpret results (as some experiments have shown that anurans and mammals can "learn" without the participation of the brain, e.g., using the brain stem or spinal cord only; see Macphail, 1982) is significant.

How we interpret comparative behavior structures its value. While many hold the idea of language being unique to humans (Deacon, 1997), other researchers find bee communication to be plastic (as noted with the ability of different geographic populations of bees to learn each other's languages (see Su et al., 2008).

But Griffin (1976) believes the dances of bees provide basic examples of symbolism. Pollio (1974) argued that there are three necessary criteria to qualify an event as a symbol: it must be representative of some other event (so bee communication qualifies here), be "freely created," and be culturally transmitted. While "freely created" is rather subjective, Griffin (1976) gives examples for bee qualification here, too, and the findings of Songkun Su et al. (2008) seem to qualify bees' "language." The central problem here is the confusion with anthropocentric ideas, as with Deacon's (1997) perspective that the peculiar complexity of human language is the central adaptive specialization of the human brain, or Hofman (2014), who believes it is the speed and amount of information that is the summation of the specialization and the cue to survival (see figure 4.2 for a comparison of different schema for human evolution). One focuses on a human mode of communication, while the other transforms the human brain into only a speedy computer (general process vs. adaptive specialization). Emery (2006) combines these anthropocentric approaches and the adaptive specialization of brain function to an understanding of avian brain evolution.

Macphail and Bolhuis (2001) argue that there are "no qualitative differences in cognition between animal species" and reaffirm Darwin's idea (1874) of differences in degree and not kind. They only admit one qualitative difference, that only humans can acquire language. But many scientists of human cognition seem to fail to realize that language is a specialized means of human quantification of experience, and that we are subject to its constraints. Our belief in its superiority is certainly a part of our reliance on it to communicate, no more or less than our inability to notice the nuances of bee language until von Frisch (1971), a process of understanding that is still imperfect (Detrain et al., 1999). Though, as Bonner (1980) notes, bees in using this language also demonstrate their ability to "teach," and European and Asiatic bees seem to be able to learn each other's language (Su et al., 2008). The evolutionary context in which this appears in bees may provide a format for understanding the appearance of language in humans (Nieh, 2010). Waggle dance location information has been shown to increase the weight of tropical forest bees and so language by spreading gathering data of food water, availability and quality may have increased

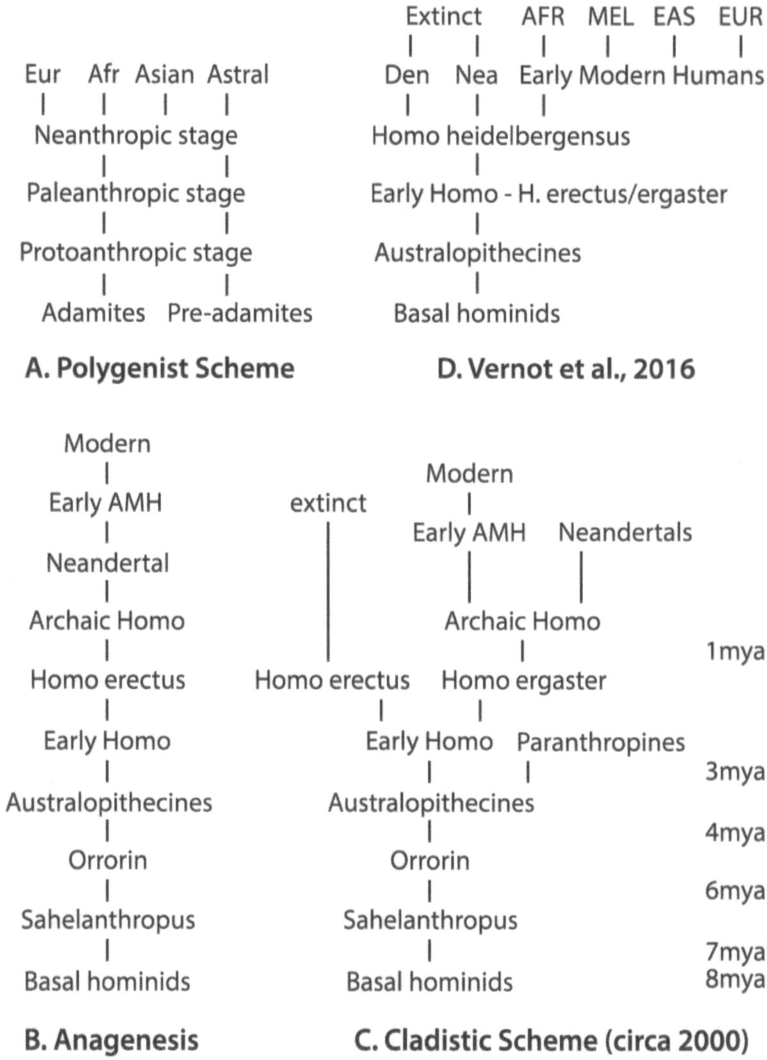

Figure 4.2 Different schema of hominid evolution. ***Source*****: Created by the author. Redrawn from Tobias (1965), Leakey (1960), Coon (1962), and Tattersall (as reproduced by the American Museum of Natural History, 2016). Figure D is derived from Vernot et al. (2016), which appears without time references but indicates an ancient division between contemporary human populations identified as European, East Asian, Melanesian, and African, with the separation for Africans as the oldest.**

survival of early hominids and may not have been associated with hunting. Prairie dog calls also show specific novelty according to unique characters and are invented ad hoc, an ability often only claimed for humans (Slobodchikoff et al., 2009).

Deacon (1996) attempts to isolate human language by appropriating Charles Sanders Peirce's distinctions of icon, index, and symbol (work published in Peirce, 2011, from 1897 and 1903). But these terms become simply tautologies as they are used interchangeably with other terms that Peirce invented, like "representan"—all of them being attempts to create distinctions between "meaning" and conditioning. An "icon" depicts a landscape; an "index" is causally linked to something. (A landscape? A smell of a skunk?) And a symbol "symbolizes" ideas that may be representations of reality as icons or an indication of something. What we are really talking about, it seems on a most basic level, is human conditioning, an ability to respond to complex sets of signals, patterned in what we call language. This can be seen with his proliferation of terms, "sinsign," "legisign" and so on (Braun, 1981). To a certain extent, Peirce is detailing conditioning, and the "learner" of language is associating constantly in a social context of signs' typical associations and contexts. We do have historical evidence of behavior becoming a "word," that is, a meaning, in this case, fuller, or felting producing a word of a different meaning over time. "Walke" or the process of fulling, felting, becomes "walk" in German. Also, the cognate in Anglo-Saxon, "wealcan" for "to roll or toss about," descended into "walk, to go from place to place: to journey, wander" (Barber, 1991; *Oxford Universal Dictionary*, 1955).

The problem with using terms like "symbol" or "abstraction" is that they enter the realm of circular definitions. They "represent" things, feelings, and meanings, and are abstractions of events that are symbolic in action and memory. All become part of a circle that produces nothing of substance but a cover for our anthropocentricism. One has to ask if the responses of birds (or other animals) to the "symbolic" matching task (Macphail, 1982) truly "represents" a meaning of mind in the animal that is parallel to that in humans when we use the term, "symbolic." Or, in an oddity task are we recognizing "differential discrimination" in stimuli that can be called "concept formation?"

Essentially, Deacon leads through Peirce's method to a complex system of conditioning and signal abbreviation to language. Peirce's scheme is no more convincing or elegant than many others produced by philosophers over the past 2,500 years. Deacon's use of it, even as a heuristic device as in "representan," seems anachronistic since it was developed before rigorous scientific experiments were designed to determine how animals structured responses to the environment and their perceptions of it, as in Rescorla's (1973) descriptions of the behavior of rats (the same finding as Farris and Breuning [1977] with red-eared turtles) of "representations of the response." Geschwind (1964), though, suggested that the origins of language might be found in cross-modal perception, for example, where an animal "feels" an object in the dark and then recognizes the same object using visual perception. The origins of language are obscure and, as Fromkin and Rodman (1978) have summarized, been a favorite subject for speculation, so much so that the Linguistic Society of Paris in 1886 "outlawed" any papers concerned with the subject. The ban was reconfirmed in 1911 and supported by the Philological Society of London. A comprehensive bibliography on language origins can be found in Hewes (1975).

CAMPS, COOKING, AND MAKING FIRE

The caveat here with the Acheulean is that, like the Oldowan, the regularity is so complete that it seems more like species-specific behavior of nest building in birds (Collias and Collias, 1984). At the time of the Levallois technique appearance, we have a number of other features: control of fire (Caldararo, 2002d; Daniau et al., 2010), first structures, grinding stones, pigment processing, and blades. Gowlett and Wrangham (2013) assert that fire was an early production of hominids, placing its first appearance at about two million years ago, arguing that evidence should be rare due to taphonomy and the identification of fire with hearths. Their first claim seems illogical, for if fire was so valuable to human survival, then it would be present wherever we find evidence of hominid activity. The lack of such evidence clearly undermines their position. The second point they make also relates to the degradation of information in sites and while it is true that archaeological sites suffer destruction, the simple importance of fire and the multiple means by which fire impresses on a physical location would survive in some residues (Caldararo, 2002d).

Brain and Sillent (1988) reported that though they believed they had discovered evidence of fire in Member 3 at Swartkrans and that remains of both *Homo erectus* and *Australopithecus robustus* are found in Members 1 and 2, they could only suggest that the ability to produce fire was discovered between Members 2 and 3 before *A. robustus* became extinct. The absence of increased use of fire in the site and in later locations at the immediate time frame would argue that either it was *A. robustus* who invented control of fire and then the technique died out with them or that it was an accidental occurrence that had no effect on the animals. The latter explanation seems incredible, as fire has been held to be such a central element to human success, so one must question this evidence. Other arguments are based on the idea that hominids could not leave Africa without fire to keep them warm. This assertion seems also without substance as many animals migrate without fire and many primates have migrated into zones colder than southern Europe or the Middle East (e.g., macaques in Japan).

Roebroeks and Villa (2011) have demonstrated that the archeological record does not support the idea that hominids in Europe had fire prior to about 400,000 BP. Yet, a recent find in South Africa at Wonderwerk Cave (Berna et al., 2012) may indicate the presence of hominids and fire in the cave. This is speculative, in that proof that humans created fire, fed it to keep it alive, and used it purposely to cook is difficult to establish. While there is ash present and fragments of bone and plant material, the concentration of fire, its use over time (necessary to establish human intervention and design) and continuous new fire is questionable. But this is only if we argue that such a powerful tool as fire would be reused immediately and over and over by subsequent members of the group. If, instead, we suppose that it was an accidental use of fire, due to a unique event (e.g., spontaneous combustion) that was utilized once and then forgotten, then the instance makes sense.

Gowlett and Wrangham (2013) also argue that fire softened foods making them more palatable, that humans cannot survive on raw foods and assert that the shorter digestive tract of modern humans must be related to this somehow. The need to increase the intake of food value can be enhanced by variations in mastication and digestion; this has taken place in birds in a dramatic fashion with the loss of teeth (Louchart and Viriot, 2011). Rotary chewing does produce a more efficient means of increasing food value (Jolly, 1970) intake by increasing the surface area of food particles so that more breakdown of food takes place and greater absorption results.

Lee and Wolpoff's (2003) assessment of cranial capacity over time shows that at about 300,000 BP, we have a median figure of just under 1,200 cc, and just after this point, we see a marked increase in brain size rising dramatically compared with the million years before. But, this is followed by a plateauing of the increase and then a definite drop in brain size. Cornelio et al. (2016) also are critical of the idea that control of fire and cooking are related to brain expansion. Theories of brain size evolution with social complexity predict just such a drop and this is demonstrated in Riveros's analysis using ant species (Riveros et al., 2012). How can we explain this "early" drop in hominid brain size during the past 100,000 years prior to the appearance of complex human social life? Striedter (2005) considers that the evidence not only supports a plateauing of brain size but a significant decrease in body size, too. Riveros et al. (2012) produce a series of charts from comparing the logs of ant brain volume, glomeruli number and antennal lobe volume to colony size.

If we chart the concept of increasing brain size in hominids with the data from Riveros et al. (2012) on ants for glomeruli (processing centers for olfactory stimuli), we see a similar drop in brain size to colony size (see figure 4.3). Brain size in ants, however, does show a drop relative to colony size, but there are outliers to this, mainly those ant species with huge colonies often described as "superorganisms." They force the curve up with a reversal of the trend in decreasing brain size. The authors suggest this is due to ecological factors, and we should note that colony size is not constant, especially for

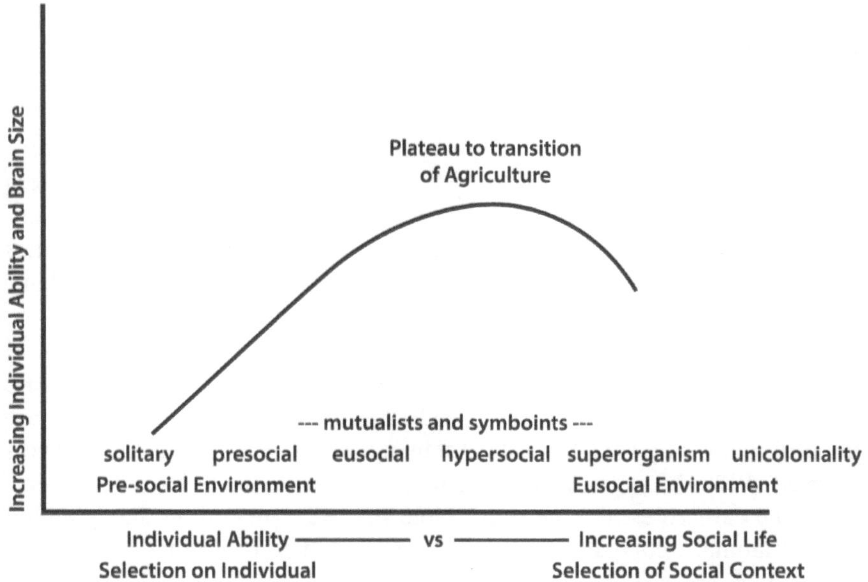

Figure 4.3 Tentative scheme of evolution of complexity in animal societies. ***Source*:** **Drawn by the author from data compiled by the author with reference to Riveros et al. (2012) on ants for glomeruli.**

termites (Roisin and Korb, 2011). But this lacks a comparison of fossil ants, though it is otherwise informative on social density. It seems obvious that we cannot clearly understand the evolutionary history of social insects, given that there are few fossil ants or termites dated before the end of the Cretaceous and only a few nests or specialized individuals (Engel et al., 2016).

But for Tobias (1971) the most significant increase in hominid brains is the leap from that of the *Australopithecines* to 1,200 cc associated with *H. erectus*. I would include specimens like Bodo at approximately 600,000 BP as Tobias attributes this period to be the greatest advances of cultural achievements, certainly evidence of controlled fire and building structures at 400,000 BP would be a safe extension. Still, this seems little to demonstrate the value of the costs of large brains. How can we justify this cost?

Usually, we find focus is placed on just a few manifestations of human cognitive behavior in the fossil record, tool making and using, treatment of the dead, site organization and art or ritual evidence. Also, new views on primate tool use and tool making are expanding our ideas of primate cognitive abilities (Haslam et al., 2009). Gowlett et al. (2012) survey the archaeological record and find a substantial difference in associative evidence, the appearance of a social brain. While McBrearty and Brooks (2000) cover the same time span and in a more comprehensive fashion, the results are quite different.

Gowlett et al. (2012) claim, as did Mithen (1996) of a functional association of domains of artifactual production that were dissociated before the creation or appearance of the social brain. A "modern integrated fashion" of thinking arrives, by an assumed "rewriting" of the brain.

While brain size slopes gradually (conforming to that of Lee and Wolpoff), the social brain is shown in figure 4.1 at the extreme right apex (as modern humans are always represented) as possessing "modern means" of intervening in the world. The problem with this measure again is that human abilities to manufacture tools is given the highest cultural value, though as I have asserted, most of this is no more significant than bird's nest construction. Still, it does not bother Dunbar and others who either promote the social brain, or the big brain thesis of human evolution, that humans (as well as no other Primate) failed to produce social structures of common effort as are plentiful in not only social insects, but many species of rodents (e.g., beavers [Gurnell, 1998; Morgan, 1868], prairie dogs [Hoogland, 1995], and a member of the carnivores, the meerkat). Note figure 2.1 regarding the evolution of the mammals, compare beavers to humans and consider that beavers and prairie dogs are rodents. When regarding these social mammals, we find that prairie dogs, for instance, not only build substantial structures, but also engage in complex social life and produce intimate individual relationships. Certainly, they should have evolved a social brain in this trajectory of group living. Meerkats also can have social groups of 20 to more than 50 members. They often stand bipedally (using the tail to support them), have binocular vision, and teach their young a variety of survival techniques (Thorton and McAuliffe, 2006). We must keep in mind that other animals, including the social insects, cichlid fish (McKaye et al., 1990) and many others, also build structures.

We have some problems in comparisons of some of these mammals. We need to know when sociality began. For example, with the beaver, fossil beavers vary in location and size. The opportunity of the finds has often ignored the context, the number of individuals, age distribution, and setting. In some cases, as in *Castoroides ohioensis*, the giant beaver of the North American Pleistocene, the average individual could be 6.2 feet and as large as 7.2 feet and was the largest beaver and rodent of the Pleistocene (Cahn, 1932). As Striedter (2005) cautions, some animals' brains have increased in size over time, while in others, the size of the body or brain has undergone reductions. *Castoroides* is a problem regarding the evolutionary history of contemporary beavers as there is another giant beaver, *Castoroides leiseyorum*, which lived earlier, and today's beaver survives in only two species, *C. canadensis* and *C. fiber* (Rybczynski, 2008). While the behavior of fossil species is most often derived from morphology, few clues survive of the sociality of the giant beavers.

Another group of interest are bats (Chiroptera). Evidence of a lack of linkage between brain size and social behavior has been published by Kerth et al. (2011), undermining the social brain hypothesis in mammals. Bats range in size over two main groups: micro and macro bats. While the complexity of sociality is substantial (Kerth, 2008) in some groups but partially divorced from kinship in some, only a few build roosting structures by modifying plant materials. That adds an aspect of extended phenotype to the social context. Female bat colony members (wild Bechstein's bat, *Myotis bechstein*) maintain social relationships over time and across fission–fusion processes of colony structure (males are solitary).

If group size became so large that grooming all members of the group became impossible to maintain cohesion, which seems to be the rationale behind Dunbar's theory, he then posits that language was an adaptation to maintain group size and cohesion. The problem here derives from a simple question, "When did this maximum group size develop to have this effect?" as there is no evidence in the fossil record that hominid groups were ever normally even half the size he theorizes as optimum (140) until the Neolithic, by which time it is usually assumed language had long been present.

Chapter Five

Smooth Brains, Convolutions, Complexity, and Ability

"The superiority of humans might simply reflect the possession of language, and the capacity for language in turn might be independent of general intelligence" (Macphail, 1982). This observation by MacPhail brings into question most behavioral research in animal cognition. Based on a general division, mammals with lissencephalic or smooth brains lacking cortical folding are small, while those with folding, gyrencephalic species, are large (Kelava et al., 2013; Zilles, 2013). The logic here was that the skull was a limiting factor to brain growth, and thus, by the evolution of folding, a brain could develop that had more surface area, which equated with more neurons and meant greater cognitive potential and intelligence. This fits well with Turner's (1890) and Orr's (1966) observations, yet studies of convolutions intensified in the latter part of the twentieth century with results that seemed as fruitful as those of phrenology. Nevertheless, the focus of explanations is on "the massive expansion and functional elaboration of the neocortex" as the factor which "underlies the advanced cognitive abilities of humans" (Geschwind and Rakic, 2013). For example, Dunbar's group size correlation fails to associate this relation with any improvement in fitness that can be demonstrated.

Is there a substantial increase in brain size before the appearance of stone tools? No; at 2.6 ma, we see little increase, nor can a substantial increase be associated with major behavioral changes like leaving Africa, as the *Homo georgicus* specimens demonstrate. They show only a small increase (about 600 cc) and are out of Africa and have tool making capacity at 1.77 ma (Vekua et al., 2002). But again, the new fossil and tool discoveries (2.8 mya at LD 350-1, see Shreeve, 2015b, and for possible tools dated to 3.3 mya at Lomikwi 3, see Thompson, 2015) undermine the big brain argument. The central question is this: How did increasing brain size improve fitness to be able to recognize 150 people, especially when early hominids most likely lived in groups of less than 50—more likely of less than 20—individuals at this time? Dunbar (2014) argues that by using the contemporary !Kung, we can project onto early hominids the value of conversation and individual recognition. He rejects the theory that language evolved to transmit technical knowledge or cultural information of ecological importance, instead seeing community bonding as the main purpose, with the usual causal, vague phrase, "... language evolved to ..." The analysis of the role of language in human evolution has been hampered by both the enormous loss of languages (Harrison, 2007) and the concentration of scholars over the past 200 years on a mythical Aryan people and the Indo-European forms of language (Anthony, 2010).

It is always problematic to define a one-to-one relationship between traits and interpreted evolutionary causes. Nevertheless, Dunbar (2014) produces a figure where he charts group size to neocortical volume and his theoretical value of "required social time" (in hours per day) using the number of fossil remains for different hominid species as group size evidence without any explanatory computation or justification for the taphonomic results. His value for hours per day is taken from grooming time in monkeys and apes. The implication from this is that beginning with *Homo rudolfensis* (some 2 ma), there has been a significant relationship between group size/social time and brain size. Comparison of this chart with data from McBrearty and Brooks (2000), reproduced here as figure 2.3, who have catalogued the cognitive products of hominids over the past five million years, we find a distinct problem. They find a unique association of innovations beginning at about 300,000 BP, which is approximately at the time of the Levalloisian technique's appearance, though we might want to push the innovations back to the Acheulean. If we compare Lee and Wolpoff's chart (figure 4.1) of the increase in brain size from the fossil evidence with that of Dunbar or Hofman, we find they do not coincide. While Hofman (1983) argues a punctuated equilibrium pattern for hominid brain evolution, his "rapid bursts" or major transitions are not associated with changes in hominid behavior that can be discerned from the fossil record. He posits a burst correlated with the appearance of the fossil hominid *H. rudolfensis* (KNM-ER 1470). The date for this fossil, about 2 mya, is too late for the appearance of Oldowan tools and too early for Acheulean tools. His second burst might fit with Acheulean tools (or early theories of the invention of control of fire, which I criticize here), but his third and the last is too late for the Levalloisian or Mousterian. Striedter (2005) argues for two leaps of increased or rapid brain size evolution, one around 1.8 or 1.6 mya and one after 400,000 BP. Neither of these fits the smoother data we have from the fossil evidence reflected in Lee and Wolpoff.

Wynn (1979) argued that by the use of Piagetian concepts of cognitive development, an analysis of hominid tools indicated a fully human consciousness by about 300,000 BP. This would push human consciousness back to about the Levalloisian and supports the chart of McBrearty and Brooks (2000). He suggested that this implied a more complex cultural life that typifies modern human cognition for hominids of this time. Wynn uses the Piagetian concepts of reversibility and conservation to assess the spatial concepts used by the hominids that manufactured the artifacts from the Isimila Prehistoric Site, Tanzania. He concludes that these artifacts required the organizational abilities of operational intelligence and that, therefore, the hominid knappers were not significantly less intelligent than modern adults. He suggests that these hominids employed operational thought that implies such cultural realms as kinship and cosmogony may have been more complex than archaeologists have had evidence to imagine.

Does the brain give us any such evidence? Do hominid endocasts? And what can the shape of the brain tell us, if the size cannot? A mixed bag results from the search for clues in the shape of brains and the idea of intelligence. In fact, Sun and Hevner (2014) find that gyrification can vary considerably between and within mammalian orders. Lissencephalic brains are found in small rodents and small Primates (e.g., some marmosets) and gyrencephalic brains are found in large rodents and large Primates. The theory proposed is that cortical folding allows the mammalian brain to grow in volume and expand the surface area despite being confined in a small skull. It would be a logical conclusion, then, that this condition would confound the idea of the brain-to–body ratio reflecting ability, as a small cranium packed with brain tissue with a small body would have the same ratio as a small cranium with a smooth "uncrowded" cranium. Herculano-Houzel et al. (2010) found a relationship between cortical folding and connectivity across mammalian species.

It has been noted that large animals have larger brains, in general, than small ones, and we note that there are substantial problems with brain-to-body ratio methods and allometry (Count, 1947; Gould, 1985; Holloway and Post, 1982). But as Striedter (2005) argues, size alone or allometry cannot distinguish aspects of behavior or intelligence, especially as allometric slopes are taxon-specific (Martin et al., 2005). (For example, figure 5.1 shows typical social behavior in prairie dogs.) It is clear that increased brain size within closely related groups, as within cartilaginous fish (the Chondrichthyes), does not mean more complex social behavior. Different pathways to encephalization have been found in various mammals, as the relationship between brain and body mass is neither simple nor a general principle (Smaers, 2012). Deaner et al. (2007) have made a convincing argument that overall brain size, and not encephalization quotient, produces the most predictive value for cognitive ability across nonhuman primates.

Sun and Hevner (2014) argue that rather than a dichotomy between gyrencephalic and lissencephalic brains, there is a continuum, and the

Figure 5.1 Prairie dogs being social and in erect posture. *Source*: Photo by Brocken Inaglory.

invention of the gyrification index (GI) defines the ratio of total neocortical surface area (including cortex "buried" in sulci) to exposed neocortical surface area.

Measurements of GI range from pure lissencephaly (GI = 1.00 in the European hedgehog to extreme gyrencephaly (GI = 5.55 in the Pacific pilot whale) and humans rank in the middle with a GI of 2.56. This is the highest of the Primates, but below many other animals, as in zebras with a GI of 2.94 and elephants at 3.81—the point being, what is the meaning of gyrencephaly? Over 100 years ago, Sir William Turner addressed this debate in a slide show of mammal brains to provide a ready comparison of brain shape using large diagrams (Turner, 1890). Turner found in his studies that there was no increase in convolution complexity in all orders nor a uniform ascending complexity or number between them and convolutions were not associated with an increase in progressive traits in other organ systems. He suggested, "As the configuration of the brain and the pattern of the convolutions have followed in each order a process of evolution characteristic of the order, the arrangement of the convolutions does not follow the same plan in the various orders." Each order appeared to have a diversity of plans of convolutions that reflected individual histories of the orders. This observation has been verified since then (Streidler, 2005).

It would seem that we should investigate the diversity of brain organization and size among the mammals, and especially between highly social members of groups like rodents (e.g., beavers and prairie dogs) and those less social. Herculano-Houzel et al. (2006) found that across six species, increased brain size was achieved by increased number and size of neurons and of non-neuronal cells but that the increase in the cerebellum was greater than in the cortex, a pattern similar with humans and other primates. Other differences become problems in comparison, for example, we find in primate brains a larger concentration of numbers of neurons in the brain than rodents of similar, or even larger, brain size (see Herculano-Houzel's images), yet as Herculano-Houzel (2009) notes, "Additionally, the so-called overdeveloped human cerebral cortex holds only 19% of all brain neurons, a fraction that is similar to that found in other mammals." So it is not just the number of neurons that counts, but also their location outside of the cortex. This emphasizes problems of the focus of many anthropologists on just the neocortex (often exempting the allocortex and cerebellum; Creutzfeldt, 1995), but even there, as Falk and Dudek (1993) and Holloway (1993) show, most analyses treat the neocortex as a homogeneous structure and assume that it just became larger ignoring the fact that its subcomponents evolved at different rates.

Barton (2012) and Deacon (1997) suggest that this concentration of neurons in the cerebellum is associated with planning, execution, complex behavior sequences in language, and tool use. Grimaldi and Manto (2012) find that cerebellar diseases impair accuracy of movements and coordination, both certainly essential in bipedalism and tool making. Also affected are oculomotor disturbances (e.g., in saccades, gaze holding, types of nystagmus and smooth following), speech deficits, ataxia of limbs and of stance and gait, and cognitive/behavioral impairments (Strupp et al., 2011). Thus, the focus on the neocortex would seem to be misplaced and complicates studies of brain size in hominids. Yet, perhaps the focus on neurons is the problem.

Sherwood et al. (2006) demonstrated the disproportionate increase in glial cell density in the human frontal cortex in comparison with other anthropoid primate species. But they also found that regions involved in specialized human cognitive functions (e.g., theory of mind, language, etc.) have not evolved differentially higher requirements for metabolic support. Though clearing metabolites and cell density and number may be related to sleep requirements that vary in mammals (Herculano-Houzel, 2015). Goldman et al. (2015) discovered that human glial progenitors are bipotential for both oligodendrocytes and astrocytes. Human astrocytes are larger and more complex than those of rodents and exhibit over triple the diameter and tenfold the number of terminal processes (Oberheim et al., 2006, 2009). This is notable as astrocytes have been shown to play vital roles in information processing within the central nervous system (Araque et al., 1999; Kang et al., 1998).

This brings up major problems with measurement of brains: where are the neurons, what are they doing and how do we compare brains across genera? Smaers and Soligo (2013) argue that brain reorganization, and not relative brain size, best characterizes anthropoid brain evolution. Specific variations in brain areas seem also to be of major importance with eusocial insects (Farris and van Dyke, 2015; O'Donnell et al., 2011); is this finding significant for human eusociality? The results of a large study of comparative animal behavior across 36 species and 567 individuals (MacLean et al., 2014) seem to support this conclusion regarding at least one factor of cognition, a behavior identified by the authors as "self-control." Brain size and dietary breadth, and not social group size, were strong predictors of species differences in self-control. Whether this quality (self-control) can be associated with any general idea of cognition or intelligence is quite another matter. Is self-control, as defined, intelligent, is "spite," as Wilson (1975) has argued, and would a comparative diagram of variation in self-control tell us anything about how cognitive abilities evolve or can be seen as a common indicator of IQ? Is the increase in brain size demonstrated by this study a result of improved self-control as increasing reproductive success, and can it be said to justify the increase in brain size? As in insect brains, which vary in size of mushroom body, complexity of calyces, number of neurons, and amount of and extent of input and output (e.g., optic lobes, medulla, etc.), in different species and genera, different behavioral adaptations appear to be the result. But how these are identified as evolutionary consequences is unclear as direct adaptations (Farris and Van Dyke, 2015).

Chapter Six

Brain Sizes, Bigness, and Neurons

In their study of African cichlid fish showing that bigger brains are associated with various factors, Gonzalez-Voyer et al. report: "Our results suggest that more complex social interactions associated with diet select for larger brains, and further that the burden of uniparental care exerts high cognitive demands on females" (Gonzalez-Voyer et al., 2009).

The problem central to the discussion of what creates big brains revolves around what Stephen J. Gould thought formed the nature of human inquiry. "Within such a view of life-as-a-whole, humans can occupy no preferred status as a pinnacle or culmination. Life has always been dominated by its bacterial mode" (Gould, 1996a). However, Hill et al. (2009) begin their inquiry with a list, a set of "key traits" of "human uniqueness," and start with "large" brains. What is a "large" brain? As I have noted, not only do we have to be concerned with our focus on ourselves as unique, but also, in making comparative studies across species, we must broaden our study beyond vertebrates to the ultrastructure, organization, and functioning of the brain in many forms (Caldararo, 2009a). Striedter (2005) explained this realization in historical context as an emphasis of focus once the problems with homology were clarified. Jerison (1973), in his comprehensive work on brains of living and fossil vertebrates, discusses some of the pitfalls. For instance, some brain cases—in reptiles, for example—are not entirely filled with "brain" tissue.

Harvey and Bennett (1983) focused our attention on metabolic activities and ability. The specific size of neurons, glial cells, and the interactions of dendrites vary (e.g., Semple et al., 2013; Shariff, 1953), and if some human brain cells are bigger than most other vertebrate neurons, then perhaps the "larger" brain idea means something other than volume alone. In fact, it is the addition of glial cells after birth that accounts for almost all the difference in cell numbers between the newborn and the adult human (Herculano-Houzel,

2009, 2012; Schoenemann, 2006), yet it seems that larger neurons are associated with larger brains across mammalian genera (Herculano-Houzel, 2013). But Roth and Dicke (2005) found that absolute numbers of neurons and their connections were a better fit with ability, yet Deaner et al. (2007), as mentioned, argue that brain size and not encephalization quotient is the significant factor. Falk (2008) makes this argument and points out another factor of interest in the variation of brain make up and organization: that women have relatively more gray matter than men. A caveat might be significant here from Darlington (2009) and Brizendine (2006), who argue that less attention has been devoted to the study of female brains than male and the sample size for reference might be skewed. At the same time, less analysis has apparently been expended to the developing female brain and its aging (see Diamond, 1988, on rat brains). But then bigger brains are not necessarily more energy efficient or computationally so (Burns et al., 2010).

Measuring brain size, either by displacement (Holloway and Post, 1982), by weight, or by volume with preserved or fossil specimens (depending on fixation, time of removal from cadaver or filling a cranium with sand or seed) results in a variety of values. Taking into consideration population variation can also produce considerable difficulties as can the age or nutritional status of individual samples. Individual entries collected by Boddy et al. (2012) reported for *Caster canadensis* are shown in table 6.1.

The median and average are about the same, 35 to 34.9, but the range is 25 to 52, while there is a cluster around 36. The sample is biased to males (9 males), and the value for the 52.21 is confusing as it is listed as for two individuals, one male and one female. It is obvious that *C. canadensis* has a larger brain than the mountain beaver, but which is also (regarding these samples) about 5 percent of the *C. canadensis* average weight. But then, the samples may contain subadults and we do not know the age status of the mountain beaver. Most of the preserved samples in the Smithsonian are fetuses and other collections have skulls without collection data (e.g., size or live weight of animal). Since we know that the mountain beaver is less social than *C. canadensis,* can we attribute this sociality to the larger species' brain size? Obviously not without a larger population of at least the mountain beaver.

In comparing the rat to another rodent, the prairie dog (*Cynomys gunnisoni*, *C. leucurus*, or *C. ludovicianus*), we find another conundrum. Values reported by Boddy et al. (2012) are shown in table 6.2.

The average weight here is 5.41 g for 13 specimens, the range from 4.0 to 7.5. The average rat brain is reported at 2.0 g (Nieuwenhuys et al., 1998), but variation is considerable. For example, the Kyoto strain in some samples can vary as much as from 1.571 to 2.064. Since these are typical rodents used to compare evolutionary trends in brain development, one can see that

Table 6.1 Findings of *Castor* (beaver) brain sizes

Brain weight (g)	*Body weight (g)*	*Number*	*Sex*
	Reported by Brody et al. (2012)		
52.21	27,670	2	M + F
25.48	4,180	1	M
29.52	5,830	1	F
36.50	17,000	1	N/A
	Mountain beaver reported by LeRoy (1987)		
1.35	465	1	N/A

At the California Academy of Sciences, I measured the volumes of the crania of eight *Castor canadensis subauritus* collected in the first two decades of the last century:

Brain weight (g)	*Specimen number*	*Number*	*Sex*
40.00	8114	1	M
39.50	7779	1	"
26.00	8018	1	"
28.00	8008	1	"
30.00	8009	1	"
44.00	8021	1	F
35.00	8017	1	M
34.90	9791	1	"

Notes: Source of Academy of Sciences samples; data compiled by the author. Body weight not available for Academy of Sciences samples; specimen number is given instead. N/A = not available.

the prairie dogs have brains that are more than twice the size. The average weight for a rat with a 2.0 g brain is 400 g; one sample has two prairie dogs at that weight with brain sizes at 7.0 g and 7.5 g. It appears that social rodents have much larger brains than the less social rats, even when body weight is considered. A caveat that should be kept in mind in terms of collected samples is that a prairie dog may vary in weight between summer and fall by a significant amount. Slobodchikoff et al. (2009) report that most animals may more than double in weight.

Slobodchikoff et al. (2009) describe prairie dogs as being able to be contrasted in the degree of sociality with Black tailed (*C. ludovicianus*) as the most social and *C. leucurus* as the least. If we put this information into the context of our data, keeping in mind the small sample size, there is a small association with increased sociality and increased brain size, as shown in table 6.3.

While prairie dogs have a variety of types of social organization of family groups, they are often highly related in specific colonies as especially in Gunnison's, but social groups do not represent family groups in terms of

Table 6.2 Brain weight for *C. gunnisoni, ludivicianus,* and *leucurus*

Brain weight	*Body weight (g)*	*Number*	*Sex*	*Species*
	Reported by Boddy et al. (2012)			
5.60	n/a	1	N/A	*gunnisoni*
5.16	786.3	2	M + F	"
5.69	992.1	2	M + F	*leucurus*
4.96	1250	1	M	*ludovicianus*
6.01	793.5	2	M + F	*gunnisoni*
7.50	392	1	M	"
7.00	504	1	N/A	"
	My samples from the Academy of Sciences			
Brain weight	*Specimen number*	*Number*	*Sex*	*Species*
4.0	7731	1	M	*gunnisoni*
5.0	9472	1	N/A	"
5.0	7792	1	"	"
4.5	7795	1	"	"
5.5	7794	1	"	*leucurus*
4.5	7793	1	"	*gunnisoni*

Notes: Source of Academy of Sciences samples; data compiled by the author. Body weight not available for Academy of Sciences samples; specimen number is given instead. N/A = not available.

Table 6.3 Brain size and sociality

Most social	→	*Least social*
C. ludovicianus	*C. gunnisoni*	*C. leucurus*
5.495	5.38	5.23
C. ludovicianus n = 8	*C. gunnisoni n* = 2	*C. leucurus n* = 3

Source: Data compiled by the author.

paternity. Gunnison's also has a considerable range of types of family group from socially monogamous to socially polygynous to multi-male/female assemblages. Usually, one form dominates in any one colony (Slobodchikoff et al., 2009). This is similar to human family organization diversity; it only lacks polyandrous unions. The complexity issue is undermined by the fact that Gunnison's, white-tailed, and Utah prairie dogs (*C. parvidens*) do not engage in social grooming.

Another example might be of interest here. The meerkat or suricate (*Suricata suricatta*) is a small carnivoran belonging to the mongoose family (*Herpestidae*). It is the only member of the genus *Suricata*. But of the mongoose family we find a number of related groups, some arboreal and some nocturnal. Most, like many carnivores, are not highly social. The meerkat and mongoose do show degrees of social life comparable to the beaver and prairie dog. The dwarf mongoose (*Helogale parvula*), which is found in groups of 8 to 30 individuals, is diurnal, matriarchal, allogroom, and highly vocal, and practices high levels of parental care in which only one dominant female reproduces (Creel et al., 1993). Another eusocial mammal is the naked mole rat (*Heterocephalus glauber*). Most mole rats are not highly social; *H. glauber* (O'Riain et al., 2008) and *Fukomys damarensis* are highly social. Burda et al. (2000) note that some species demonstrate a division of labor with cooperative food gathering, digging, cleaning, and brood care. Variations exist in reproduction in males and females in different groups and a high number of colony members fail to breed, giving mole rats a higher degree of relatedness than some social insects (Faulkes and Bennett, 2001).

So, while we may have some association of brain size and eusociality, the case of the mole rats seems to argue that very small brains (given our only reported sample of *H. glauber* has a brain weight of 0.52 and body weight of 60.8 g) sociality and complexity is quite possible. For very small brains, Boddy et al. (2012) used Isler and van Schaik's (2006) reduction factor of 0.59 g, which may not be appropriate. So, perhaps looking at the distribution of neurons would be of more help. The literature on the *Carnivora* supports the idea that increased sociality is associated with increased encephalization, yet when Finarelli and Flynn added in fossil samples to the database, they found that there no longer appeared to be a causal relationship between sociality and encephalization for this clade (Finarelli and Flynn, 2009).

While the number of neurons varies in a discrete vertical cylinder of cortical tissue across species, the neocortex of all species contains a set of elements both similar and comparable, while control of the growth and development is under the control of a number of genes shared by mammals (Striedter, 2005; Sun and Hevner, 2014). However, in a comprehensive analysis, comparing the human brain to that of other mammals, considering different cellular scaling rules that apply, it was found that the human brain is not exceptional in its cellular composition and contains as many neuronal and non-neuronal cells as would be expected of a primate brain of its size, and the number of neurons is similar to that found in other mammals (Herculano-Houzel, 2009). Cartmill (1990) came to a similar conclusion in criticizing the various methods of comparing brains of different species to that of humans and the required assumptions applied to the data by different scientists to be able to call the human brain unique. It is built very economically.

While arguments over scaling failed to resolve issues of the size of the brain, Holloway focused on relationships between brain nuclei, fiber tract, and neuroreceptor sites in a growing concentration of attention on the idea of a reorganization process in the human brain that could be responsible for human cognition (Holloway, 1974, 1979; Holloway et al., 2004). Some of this contrast was seen as a false dichotomy (Falk, 2007b; Gould, 1985), but has regained interest recently (Striedter, 2005).

Chapter Seven

A Brain of Two Parts

Cortex versus Cerebellum

We are overwhelmed by the size of things, and the sizes of parts of our brain have fixated the attention of scientists while what comes out of that brain and its parts is often ignored or idealized. As Tinbergen remarked, "The consummatory act is relatively simple; at its most complex, it is a chain of reactions, each of which may be a simultaneous combination of a taxis and a fixed pattern" (Tinbergen, 1951).

Herculano-Houzel (2009) describes the history of regarding the human brain as special as confused with the idea of a large cerebral cortex. She notes that brain size has become a proxy for the number of neurons in the brain, the cells that do the work. Yet she also notes that the human cerebral cortex only holds 19 percent of all brain neurons. But what of the other 81 percent of the neurons, where are they and what do they do?

Weidenreich (1949) demonstrated that in terms of brain-to-body ratio, humans do not have the highest ratio. Whales have a ratio of 1 g brain to 8,500 g of body; man, 1 g brain to 44 g of body weight; dwarf monkey (marmoset), 1 g brain to 27 g body weight; and capuchins, 1 g brain to 17.5 g body weight. But as Striedter (2005) cautions, that it is difficult to discern evolutionary trends from brain-to-body ratios as some species may have undergone a decrease in brain size relative to body size or vice-a-versa as may have taken place in the evolution of beavers.

There has long been a belief that human brains were the only ones with laterality (Halpern et al., 2005); rather there is considerable variation, with human brains more asymmetric than many (Rogers and Andrews, 2002). The belief has been shown to be false across many animal genera; in fact, laterality seems to be associated with social behavior. It is remarkable how little comparative study was done in the past before people made generalizations about human uniqueness. However, recent studies using fMRI appear to undermine concepts of lateralization based on human lesion studies (Huth et al., 2016). Also, there is apparent lateralization related to gender (Falk, 1987; Holloway, 1996).

We expect differences among mammals, but is there a uniqueness? What defines uniqueness? Categories are problems; humans have been identifying the natural world for millennia with our own perception of reality and creating linguistic and cognitive divisions (e.g., species), yet even to say this is an undertaking in anthrocentricism and certainly ethnocentrism (Harrison, 2007). Striedter (2005) begins his discussion of the evolution of brains with the statement that he will limit his analysis to vertebrates, yet he does not give a definition of a vertebrate and what is included or excluded and why. It once was thought that birds in general had small brains compared to mammals and, therefore, had to be "stupid." This was because we "saw" birds in terms of mammals, and this was natural since, well, we are mammals. It was then realized as behavioral research and comparative anatomy advanced, that birds had behavioral capacities far exceeding what would be expected given their small cortical mass. Experimental research then showed that in evolution, bird brains underwent a massive development of a different part of the primitive reptilian brain than mammals. The Wulst was the tissue that formed in birds to function as the higher processing area as opposed to the neocortex in mammals (Stettner and Matyniak, 1968), though it generally functions processing visual information. Together with the dorsal ventricular ridge, these two regions of the bird brain can be said to be somewhat homologous to the mammalian neocortex (or vice versa, see Striedter, 2005). However, we are again fooled by our preference for size, as Olkowicz et al. (2016) have found that the brains of parrots and songbirds contain, on average, twice as many neurons as primate brains of the same mass, indicating higher neuron packing densities than mammalian brains. They also discovered in large-brained corvids and parrots, forebrain neuron counts equal to or greater than primates with much larger brains.

When we consider the range of reaction and capacity in animals across the major classifications, we find a general grading in ability (Macphail, 1982, 1987). This seems true even in the divisions recently constructed on the basis of different contexts, such as in social versus ecological intelligence (Steklis, 1999). Critics of this perspective utilize experimental designs that impute to other animals' motivations that are subjective with results tallied as "apparent" outcomes interpreted by the researchers as if they could examine the thought processes of the subject animals objectively (Premack, 2007). It is remarkable to note how often researchers argue that nonhuman animals do not think, when they mean that they do not communicate their thoughts. It is as if the researcher could "read" the minds of nonhuman animals. Donald R. Griffin addresses this conflict in his 1976 book, *The Question of Animal Awareness*. John R. Searle elaborated on this issue in his "Animal Minds" article (Searle, 1994) showing that demonstration and interpretation of what animals and humans "think" is more of a philosophic question than a scientific one, even though we can chart movements with MRI and make assumptions about intentionality (Bekoff, 2013). Interpreting animal behavior, as when a cat or dog travels across the country to return to its former home after an accidental trip (http://www.brainz.org/15-pets-who-returned-home-remarkable-distances/), is problematic as most cases have little in the way of proof that the animal made the trip at all, and in others, intentions are difficult to discern as the discovery of the route seems supernatural. A more important question is what does it mean to be human? And, why do we find such ideas of intentionality in animals so remarkable?

According to Tobias (1971) and Holloway (1996), we can compare the trends in brain size development in the appearance of certain fossils, so let's review evidence (see tables 7.1 and 7.2).

Table 7.1 Fossils, brain sizes, and time

Fossil	*Time frame*	*Brain size (cc)*
Australopithecus afarensis	4 my	350–400
New Stone tools at Dakika	3.8 my	
Homo habilis (KNM-ER 1813)	2 ma	612
Homo rudolfensis (KNM-ER 1470)	1.8 ma	781
Homo ergaster (KNM-ER 3733)	1.3 ma	850
Homo erectus (Zhoukoudian 1927)	600k BP	994
Bodo	600k BP	1,100
Kabwe (Broken Hill)	300k BP	1,283
Omo Kibish	195k BP	1,330
Nahal Amud, Israel	45k BP	1,740
Le Moustier	40k BP	1,750
Homo floresiensis	94k to 13k BP?	405
Median modern human	0 BP	1,350

Source: Data compiled by the author. All dates are approximate.

Table 7.2 Time frame and associated cultural products

2.6 ma	*1.6 ma*	*400k BP*	*20k BP*	*10k BP*	*7k BP*
Oldowan tools	Acheulean tools	Levallois tools	Atlatl	Domestication	Written language
		Fire, clothing	Bow and arrow	War, towns, boats	
		Structures, fire			
		Hardened spears			

Source: Data compiled by the author.

NANOCELPHALIC DWARVES, *HOMO FLORESIENSIS*, AND WHAT IT MEANS TO BE HUMAN

There have been many studies produced in the past 200 years that focused attention on brain size and intelligence. Most of these were the result of prejudice as can be seen in one of the earliest examples, *Crania Americana*, by Samuel George Morton. Some were more or less transparent in intent given the poor research designs; others exploited new methodologies that caused confusion in scientists and the public when published. Ashley Montagu reviewed most of these in his book *Man's Most Dangerous Myth: The Fallacy of Race*. The same is true of studies of IQ and inheritance of ability; Montagu critiques those authors who attempted to give scientific cover for racist ideas of intelligence, and Steven J. Gould gives a much more detailed report on this phenomenon where he debunks a number of proposals of inherited ability and IQ in his book, *The Mismeasure of Man*, which appeared in 1996 with an expanded section to refute the claims of the authors of *The Bell Curve*. Ideas of uniformity of brain size and cranium size were seldom recognized to vary with age across species and genera, as in the case of many mammals (see skunk crania in Elbroch, 2006). Explorations of the stability of IQ over time, as that by McClearn et al. (1997), demonstrate the remarkable limits of increasing ability over time, and whether this is a genetic barrier or a cultural one is yet to be approached, though Kamin and Goldberger (2002) have questioned the methods of these studies.

Ideas of brain size evolution have been varied, though some problems have developed due to nationalist or racist ideologies as when the Piltdown Hoax confused anthropological research, creating false information that very early hominids had brains as large as contemporary humans. Kappelman (1996) clarified a number of other problems, for example, where it appeared that Archaic hominids (*Homo heidelbergensis* and Neanderthals) had significantly larger bodies than contemporary humans or early modern *Homo sapiens sapiens*. Ideas of an association between large brain size and intelligence have also been undermined (Schoenenman et al., 2000) due to studies that have confounded between-family environmental influences with direct genetic/biological influences.

Both Gould and Montagu address the inherent vice of intelligence tests and such assessments in general. Subjectivity seems to dominate the field as both have demonstrated. Definitions of ability vary with research assumptions, and there has been a spirited debate in the field for decades over these subjective parameters, a paper by Nathan Brody ("g and the one-many problem: is one enough?" Brody, 2000) attempts to produce comparative parameters by studying various animals. His conclusions may overstate the case for measurement, but the discussion outlines the issues in interpretation. Some efforts have centered on baselines using chemically produced "microcephalic" mouse models and inbred lines of mice (Anderson, 2000).

The issue of microcephaly has been raised several times in the past 100 years as a means of understanding the evolution of the brain and intelligence in humans. I think that brings up a discussion of what it means to be human. The idea of big brains as a necessary component of anatomically modern humans is a proposition. The fact that Neanderthals had the biggest brains of hominids creates contradictory explanations. The idea of the recent find on the island of Flores in Indonesia is a case in point. Some paleoanthropologists consider the specimen, now dubbed *Homo floresiensis,* to be a new species (Falk et al., 2005); others think it to be a microcephalic dwarf (Martin, 2006). This idea is of interest to this discussion, also, due to the small brain and body size yet obvious evidence of considerable cognitive ability.

Nanocephalic/microcephalic dwarfs (the term varies in the literature) can learn language and are capable of a number of direction-led actions (Green et al., 1995). A comprehensive analysis appears in Helmut P.G. Seckel's book *Bird-headed Dwarfs* (Seckel, 1960). Professor Martin and others chose to identify the *H. floresiensis* specimen as a microcephalic due to the characteristic mandible deformation. When you look at the fossil and then the examples in Seckel, you see the point. There is also the issue of rapid ageing, which can be associated with the *H. floresiensis* specimen as an example of a nanocephalic/microcephalic dwarf. More interesting is the evidence of other individuals appearing in *Smith's Recognizable Patterns of Human Malformation* by Kenneth Lyons Jones (1988). Here, we have a number of other features, which also show similarity with the *H. floresiensis*; in one example, two children, brother and sister, are shown side by side. Remember, they are the size of large Cebus monkeys. The cerebrums of both the boy and the girl are unusual. The cerebrum shows a "simple primitive convolutional pattern resembling that of a chimpanzee," according to Smith. They have only 11 pairs of ribs apiece; they cannot straighten their legs fully; like many "bird-headed dwarfs," they have displaced hips. Others have displaced elbows. Yet, given all this, some have lived to 75 years of age.

Two studies by paleoneurologist Dean Falk and her colleagues (2005, 2007) rejected this possibility. Falk's (Falk et al., 2005) arguments have been rejected by Martin et al. (2006) and Jacob et al. (2006), and defended by Morwood (2005) and Argue et al. (2006). Weston and Lister (2009) have also produced comparisons with other mammals and dwarfism with supporting evidence for *H. floresiensis*. What is most interesting in this find of *H. floresiensis* are the tools that appear to have been made by these people. While this particular find may be a microcephalic dwarf of a normal *H. sapiens* population, the idea that an individual with such a small brain could make tools at all, and in this case fairly complex tools, undermines the theoretical frameworks of brain evolution and cognitive requirements for human cultural complexity.

A more recent publication focuses on the dentition, and some support for microcephaly was published by Regen et al. (2010). It must be kept in mind that this condition is regarded as an autosomal recessive trait, but the fact that there seems to be some variety of outcome has caused some researchers to argue that there are type I and type II forms. In fact, C.G. Woods (2004) considers there to be at least six genes involved in the various contemporary forms. How these genes interact is unknown. The genetic evidence of contemporary examples does not seem to be well characterized as yet and it seems premature to conjecture about the evolution of the genes involved until we have more precise information of the living varieties. Richardson (2011) describes the some 30-odd genes that seem significant in the study of brain evolution and cognition and cautions concerning the difficulties of relating single gene functions to performance. She also critiques some recent interpretations of the function of these genes and potential implications on IQ and race.

In addition to the gene variants, microcephaly can be induced by prenatal infections and mother's alcohol consumption, mimicking the action of Methylazoxymethanol in mice (Anderson, 1994, 2000; Ornoy and Ergaz, 2010).

NECESSITY OF THE SOCIAL CONTEXT TO BE HUMAN

We have to keep in mind that brain development in newborns has specific survival strategy elements (Pagel and Harvey, 1988); it is true that humans have only about 25 percent of their brains developed at birth (Eisert et al., 2013). This provides the developing brain with a social environment to initiate the context of language as a basis for a human consciousness, or what Chomsky (1995) called generative grammar. Without that context, language is unlikely to develop, the fantasy of wild children raised by animals or by demented relatives aside (Bettelheim, 1959; Singh and Zingg, 1942). Bettelheim attributes most cases that have any verified history to pathological individuals abandoned by relatives, and Singh and Zingg's examples are all unverified cases based on hearsay (Ashley Montagu, 1943; Brown, 1958). Like the famous case of Kasper Hauser, fact is rare but mystery large (Heydenreuter, 2003; Striedinger, 1933). Salimbene di Adam wrote in his Chronicles of experiments carried out by Holy Roman Emperor Frederick II to isolate infants to see if they would speak an original language, but instead they spoke nothing.

The same experiment was supposed to have been carried out by the Mughal Emperor Akbar (Abul Fazal, 1993).

Even deaf and partially deaf children, exposed to linguistic frameworks communicated from speaking parents without manual linguistic aids such as Braille, acquire basic communication from the language and a rich visual environment (Goldin-Meadow et al., 1984). Many such individuals in the recent past, and those with dysphasia, were placed in institutions as their parents and caregivers considered them incapable of language or intelligence. But given that this condition varies greatly, from developmental dysphasia to that caused by stroke and injury, comparisons of reports in the literature are fraught with uncertainty. Dr. John Bulwer published in 1648 a treatise on teaching those without speech or hearing to communicate, though earlier claims were made of success. Likewise, some cognitive tests were made of these children and the results indicated low linguistic ability but normal performance on "nonverbal" intelligence tests (Cromer, 1978). Macphail (1982) suggests rejection of these children by their parents due to their behavioral deficits probably enhanced their identification as abnormal.

Helen Keller was blind, deaf and mute but was exposed to language as a child before the age of two. This case was similar to that of Laura Bridgman and a number of others (Widd, 1880). There are no verifiable cases of language developing without a social context.

Chapter Eight

The Future of the Human Brain

We have examined many of the recent theories that attempt to explain the big brain in humans. But as Levi-Strauss (1969) noted, "In science there are no final truths." The late appearance of tools long after bipedalism has become habitual, the small size of the brain long after the invention of tool making and the slow increase in brain size from two million years ago to the Neolithic all fail to provide a logical relationship. The small sizes of human groups over the past seven million years with only recent (in terms of 10,000 years BP) indications of eusociality argue against the social brain theory. It appears that the body size of humans has decreased in the past 50,000 years, which gives a false impression of brain size increase, and yet, there appears also to be a real reduction in brain size, which parallels that in domesticated animals.

It seems likely that brain reorganization, modifications in neuron structure, and dendritic connections have some relationship to social life, and we see parallels in increased brain size in beavers and prairie dogs over other non-eusocial rodents. The data supporting that argument in this book may be an artifact of measurement or small sample size. A larger study with preserved brains is needed. The data on *Carnivora* undermined this idea.

The review of comparative brain neurology and anatomy tends to force a conclusion that brain size increase was related to competition for food by hominids that intensified during the Pliocene with decreasing food supply. Tool making may have been a component in a suite of adaptations that selected for greater variability in food identification, memory of location, predator avoidance (as in brain differences between spider monkeys and howler monkeys), and the addition of hunting skills. The increase in brain size appears to have reached an apex in the Upper Paleolithic, at which maximum benefits over deficits was met. After this, with the invention of sedentism, human life began to be organized with more density. This would support the idea that the expansion of the brain after 2.5 mya was due to the increased efficiency in controlling movements, especially fine movements in tool making and tool using, producing cognitive field capabilities for analysis of traces of information relating to identifying foods, tracking animals, and locating and memorizing water sources. This would be also consistent across orders with the parallel increase in hunting skills in cephalopods and spiders, especially cooperative hunting and securing food for later distribution. Here, we have the framework for later social complexity.

The Neolithic was constructed as a niche peculiar to human social life and communication became more significant, as well as the ability of the individual to adapt to increasing demands on conformity and rigid social conditioning schedules necessary for group life in sedentary communities based on domestication of plant and animal life. Nevertheless, it seems logical, following the example of insects and the appearance of mushroom bodies, or in social spiders and large-brained cephalopods, that the rewards of greater food collection and hunting along with better defense, especially the protection of kills cooperatively made, brood protection and care, led to the big brain. Deacon (1997), however, links hunting with parental care and sexual selection, arguing that early hominid females would have had a near-impossible time hunting and protecting their young. So, he creates a suite of selective pressures around the idea of hunting and monogamy as a driving force for brain evolution. This is curious and not supported by the behavior of other predators, like felids and bears whose females do hunt and care for their young alone (Broch et al., 2015; Lopex-Bao et al., 2014). The random loss of their young to other predators certainly acts to hold their population in control of the available prey.

Social spiders' behavior is often termed "quasi-social" (Brach, 1977) given their cooperative hunting. Yet, for example, in *Anelosimus studiosus*, there is cooperative web construction, capture and utilization of prey, and indiscriminate brood care with colony permanence.

Food production and sedentism led to the opposite trend. This established the subjugation of humans to their own domestication, which has continued to this day. It is of interest in this regard that human brains have slightly reduced in size over the past 50,000 years. This might be a response to reorganization of the brain, corrections from cold adaptation in some groups or the effects of self-domestication. We know that one difference between wild boar and domesticated pigs is the reduction in size of the brain of domesticated pigs and that if they become wild, it is not reversed (Kruska, 2007). There is perhaps a similar trend in social insects, with an association between decrease in brain size with increase in population density (Riveros et al., 2012). Yet, distraction of domesticated animals, especially in dense population conditions, reduces damage and makes herds easier to control, feed and cull (Price,

2008). Distraction in human society may play a similar role, note today's ever-present electronic device, from smart phone to laptop. Domestication and the elimination of predators are associated with a decrease in brain size (Palombo et al., 2007). Since humans after about 50,000 BP had tools, techniques, and organization to remove most sources of predation on them and developed self-domestication with urban life and food production after 10,000 BP, one might argue that the effects of these two trends could explain the reduction in human brain size. Self-domestication could be equated with Alexander's (1989) ideas of the role of "self-deception." On close inspection, however, Alexander seems to be speaking more of Durkheim's (1915) representations and their context.

We might consider this information about decreasing brain size as it affects the daily lives of social individuals. There have been patterns of repression that had substantial effects on ability in human society, as at the end of the Roman Republic and continued into the Empire, where wave after wave of military murders swept across the empire, striking down members of the aristocracy and the most able of the plebs. Rostovtzeff (1926) reports that this terror became institutionalized and focused on the intelligentsia in particular after the second century AD; the effects of such a campaign against ability can be seen in the collapse of the empire's institutions. This was the focus of Tobias' (1971) rather negative view of man's future as he assessed sympathetically Eiseley's (1958) conclusions. One might argue, too, that Armstrong's (1989) idea of "runaway" competition can function as an impediment to cognitive progress by the selective elimination of those of "different" ways of thinking, of thoughtful competitors or clever opponents to produce the serf/master or caste limits of some historical social organizations (Dumont, 1966). We might then suggest that the purpose of language is as an adjunct to economy in complex society (bees and humans, certainly), but it also enhances recruitment efficiency and unit solidarity. Intelligence then becomes a function of complexity and its uniformity a component of integration. Thus in humans, repression, Nazi holocaust, ethnic pogroms, religious inquisitions, and other mass psychogenic diseases (Caldararo, 2012b; e.g., Salem Witch Trial, or searching for witches among the Hopi) all act to enforce a uniformity of intelligence, creating upper and lower limits.

Yet, the invention of tools and the character of the human hand (Napier and Tuttle, 1993) should not be overlooked, especially in light of our comments on the cerebellum. Tools allowed and accelerated the elimination of all animals that were main competitors with hominids, and all that were predators to hominids, faster and more efficiently than any other animal has achieved, which we can detect from the fossil record. Tools expanded the "expanded phenotype" sufficiently to delay the development of complex society seen in other animals as these social structures, while also extended phenotype, do not appear to be as effective nor achieve the level of dominance that tools have provided hominids. Other animals can recruit others, communicate plans, and project these into time and space, yet tools have provided clear avenues for the extermination of all predators, access to a wider variety and amounts of resources, and durability and variety of built environment.

Other explanations have shifted attention to changes in brain areas rather than absolute brain size (Schoenemann, 2006). Nevertheless, increasing brain size in eusocial mammals seems to differ from that in humans in that, while these species all live in highly social environments, as individuals they are still largely given to individual behavior for survival, especially as they do not have domesticated plants and animals to either produce food for them or provide energy in the form of labor. Yet, some activities are organized and there is some degree of food sharing. Therefore, compared with non-eusocial mammals, say rats and felids, their individual source of food and energy is largely self-motivated even though prairie dogs experience considerable cooperative interactions and highly social structured behaviors. The question is how much social behavior matters, or what kind? This argument seems applicable to large-brained cetaceans, birds, and cephalopods. The increase in beaver and prairie-dog brains would parallel the increase in hominid brains prior to the discovery domestication of plants and animals and the invention of human self-domestication. Human self-domestication would then be the most logical factor in the reduction in human brain size along with the reduction in predation, perhaps in combination with the well-known association of reduced body size in relation to increased population density where energy use appears independent (Damuth, 2007).

As members of large, dense populations of complex societies, human individual behavior has become dependent on social processes to survive requiring less rigorous individual abilities, and this has been intensified with mass media and standardized group education institutions as well as religion that instills group identity and identification with symbolic entities that theoretically govern human agency as mediated by leaders in political and religious spheres. Emphasis on individual decision-making is reduced and this is reinforced recently with electronic media and neuromarketing techniques for religious, political, and commercial motivation. Malinowski (1935) equated advertising with magic. Anyone exposed to the images of advertising can recognize the illogical, magical, and irrational nature of much that we see on TV and the Internet as advertising. Studies of how individuals make decisions and what information and methods they use reflect socially determined options within a collective habitat of values according to Etzioni (1988). Those who control media (and questions regarding its concentration) are increasingly manipulating this collective habitat, and questions have been raised about whether the Internet has significantly affected it as the companies that dominate it by their outlets have mushroomed (Noam, 2009). There have been intelligent and informed peaceful rebellions to corruption and inequality, as in California during the Progressive Period when professionals worked with craftsmen and workingmen and women to overthrow the political control of a small group of wealthy individuals (Starr, 1985). However, the transformation of information and documentation has been so comprehensive that it is unlikely this could occur again, as O'Neil (2016) describes for the growth of communication and information technology. The ability to hide corruption and exploitation seems to grow at a remarkable pace—as in the 2016 Well Fargo scam produced not by criminal outside hackers but by the inside management and employees (Corkery, 2016)—and in complex algorithms.

Stith and Vigil (2016) argue that inequalities like slavery (e.g., feudalism) and discrimination can produce actual opportunity gains, and moral constraints on such institutional social asymmetries act as opportunity costs undermining any comparison of human society with eusocial insects where castes and gender limitations are present. This idea is unsupported by evidence, as Adam Smith (1776) argued; class and discrimination produced corruption, and advantage often advanced the inept, but class society also often results in cycles of destructive ritual conflict (Caldararo, 2013). They also assert that controls like those on usury or organ sales that would increase productivity do not exist in ant society. This ignores the fact that laws and sanctions do not eliminate such behaviors: we still have illegal banking behavior like those of the credit crisis, loan sharks, and the sale of organs (Scheper-Hughes, 2000) illegally, as well as child "adoptions" and surrogate births that are thinly disguised sales.

In fact, many of the commentaries to the Gowdy and Krall (2016) article harp on the idea that humans are so exceptional, their societies cannot be compared to those of insects. But perhaps these qualities and abilities, "thinking," making airplanes and computers, and language are all simply the noise of scale, the "hum" of the hive. Do they really matter if the march to superorganism continues and the interconnectedness of humans comes to function like that of insects in their societies? On a related note, Cipolla (1976) implied that much of human behavior is stupid, ignorant, and often characterized by folly and delusion. A similar assessment is to be found in Montagu and Darling (1970).

Animal studies can teach us much of the basis of the evolution of brains. As Macphail (1982) argues, however, severing an area of the brain from other portions and then deciding what has happened to functionality is not the same as knowing what is going on in that animal's mind. The fact is we

	Building	Food sharing	Brood care	Agriculture	Husbandry	Protection	Division of labor	Grooming	Coll. hunt/gather	Population control	Language	Tool-making
Prairie dogs	●	●	●			●	●	●	●	●		
Beavers	●	●	●			●	●	●	●			
Humans	●	●	●	●	●	●	●	●	●	●	●	●
Fish	●		●			●	●	●				
Dolphins		●	●			●	●	●	●		●	●
Naked mole rats	●	●	●			●	●	●	●	●		
Bats		●	●			●	●	●	●			
Birds	●	●	●			●	●	●	●		●	●
Bees	●	●	●			●	●	●	●	●	●	
Wasps	●	●	●			●	●	●	●	●		
Ants	●	●	●	●	●	●	●	●	●	●		
Termites	●	●	●	●	●	●	●	●	●	●		
Social shrimp		●	●			●	●	●	●	●		
Squid (cephalopods)		●	●			●	●		●			●
Beetles	●	●	●	●		●	●	●				
Social spiders	●	●	●			●	●		●			

Figure 8.1 Distribution of traits of eusociality. ***Source*****: Created by the author.**

do not know what the words we use—cognition, mind, language, meaning, and the like—represent, as they are our creations based on assumptions of our own interaction with reality, which is mediated by culture and conditioning. Our understanding of what happens in our own minds is imprecise at best, which is why treatments and assessments of illness in psychology are so frustratingly poor in outcomes of cure (Laing, 1960). These problems, often leading to conflict between people who love, end in divorce, those between families and communities end in violence, and those between nations end in war and brutal massacres. The nature of reality is at risk, and yet Freud (1966) provided a valuable explanation for the experiences people had regarding its nature when he examined dreams and concluded that they contained the residues of daily life mixed with the distorted memories of childhood, when the developing brain of the child is exposed to images it could not process and were not linked to linguistic contexts of meaning. These "shadows of reality" haunt dreams as frightening, strange, and confused shades that in religious contexts have been explained as worlds of spirits and gods. The lack of certainty is the central problem that we fail to achieve and yet we act on the fragments of probability as if it were reality.

We might argue, in conclusion to the issue of why we have big brains, that:

1. The ecological theory associated with cooperative collecting/hunting has some relevance given the parallels with eusocial insects, spiders, fish, dolphins, beavers, and prairie dogs.
2. In hominid evolution, increased pressure on the cerebellum—for coordination over more agile bipedalism (e.g., running); intense, concentrated use of eye-hand coordination for tool making/using for specific purposes; and perhaps language organization (Deacon, 1997; De Smet et al., 2013)—was significant.
3. That hunting/collecting and predator evasion also were pressures for increased processing in communication demands.

These three areas led to the preconditions for eusociality in hominids after 10,000 BP, including sedentism and domestication of plants and animals. Then we can see that more effective collecting on a cooperative scale leads to resources for sedentism to institutions for complexity and then specialization and the potential for eusociality (see figure 8.1).

Part II

HISTORY OF A GENUS AND THE EVOLUTION OF SOCIETY

Chapter Nine

Anthropocentric or Indifferent Universe?

> A description of a cosmology where man is not the center is given by Crapo. A people have a god who when they die dries their souls over tree branches. When the souls dry the god wraps them up and smokes them to have pleasure.
>
> (Anonymous)

We evolved under conditions of certain temperature and pressure. The functions of our cells, the incorporation of proteins, minerals, water, the pumping of the heart, the movement of synapses, are all adaptations to recent earth conditions of the past billion or so years. Would we still produce a human cognition if these conditions no longer existed, for example, in outer space? What is the reality we conceive of other than the recognition of the structure of these conditions? Muller (1892) talks of the "natural religions" as being reflections of perception, of good and evil, and the nature of the social structure as creating the god of Durkheim (1915). The nature of our cognition has allowed us to see the world in a certain form, and whether that form is reality depends on how our senses had evolved to interpret physical features of the environment. We then respond with behaviors that seem appropriate, and if they are, we survive. At present, our abilities have produced things like more accurate atom bombs, and one has to wonder how intelligent that design is for a future survival. De Waal Malefijt (1968) discusses the variations in the idea of a nonanthropocentric universe with several examples from different peoples.

Primates came into existence after a dramatic increase in temperature at the beginning of the Eocene, called by paleoecologists the Paleocene–Eocene Thermal Maximum after about 55.8 million years ago (Woodburne et al., 2009). This was followed by a period of stasis for about 5 million years, and then another dramatic climate episode, the Early Eocene Climatic Optimum, took place between 53 million and 50 million years ago. Temperature increased to the highest prolonged Cenozoic ocean temperature and mean land temperatures, causing increased rainfall and "promoting a major increase in floral diversity and habitat complexity" (Woodburne et al., 2009). Avian and mammalian metabolic needs are three to six times as high per unit of body mass as reptiles and the diversification of placental mammals occurred between 65 and 100 million years ago, corresponding to a period of high and stable oxygen levels in the atmosphere (Falkowski et al., 2005). Under these conditions, Primates originated within the expanding and diversifying new species of plants, especially forests. This was followed by a collapse of conditions from 50 million to 47 million years ago, resulting in major faunal diversity loss called the Bridgerian Crash. Variations in temperature and rainfall continued to the late Eocene and effected such changes as the decline of the Adapoidea and the rise of the anthropoids.

In like fashion, the conditions under which our genus, Homo, came into existence resemble the origins of our Order, Primates. From about 900,000 BC until about 10,000 BC, the earth experienced a number of remarkable and dramatic changes in temperature and humidity, but in thiscase, we can describe these as warm and wet periods interspersed with long glacial ages. Most warm periods (interglacials) lasted around 10,000 years or so, while glacial periods were much longer, some 100,000 years (Klein, 1989). So whereas our Order arose in warmth and in wet conditions, varied by dry or stable periods, our genus suffered under mostly extreme cold and dry conditions, varying with warm and wet ones. This cycling affected all life, plants, and animals, selecting for durability and adaptability. Survival of hominids in this time can be seen dependent on a more robust physiology of the skull and postcranial skeleton. Life was harsh for most of the time and some of the injuries we find with mid-Pleistocene hominids, seen most evidently in Neanderthals, are similar to those of today's rodeo riders (Trinkaus, 2012). Yet, the cranial capacity continued to increase, though slowly indicating that cognitive ability must have depended on plasticity of behavior and social support. To a certain extent, the cranial architecture demonstrates this adaptability in that the face in most mid-Pleistocene hominids like Kabwe (considered today, *Homo heidelbergensis*) in Africa, or Petralona in Europe or Dali in China, all show prognathic or projecting faces (Day, 1986). Such projection of the face can provide for a multitude of sinus cavities where mucus membranes can give off moisture to dry air and collect debris from it, or warm cold air, in both cases, protecting the lungs from damage (Gamble, 1993). I would also assume a substantial amount of sinus surface area andmucus membranes, and large noses, would account for the faces of African specimens like Kabwe and Bodo (Tucci, 2016).

The ways that people have explained the conditions of life and its origins have produced the stories of religion and magic. These creative myths and legends involve culture heroes and gods, yet they are no less complicated or inspiring than science's current Big Bang Theory—except that, in the case of the Big Bang Theory, it conforms to the physical examination of conditions in the universe, can explain much more logically, and is testable as well as being verified in most details. Scale and familiarity with the information one is presented with are essential for understanding any representation of a cosmology. For example, we immediately class similar distributions of information in a familiar context, as in figure 9.1 at the beginning of this chapter, which shows mold growing on an emulsion. Most people are so used to imagining the appearance of our universe that, shown this image of mold, they see it as an image of the universe and attempt to identify star systems and constellations.

The paramecium swims in solutions of chemicals, producing what we perceive as a "trial and error" type of movement. In the early twentieth century, observations of this behavior led to it being called "phobo-taxis," or fear movement, by Kuhn. Since there was no constant relation between the new direction of movement and the direction of the stimulus (change of chemical composition, heat or cold, solid object, etc.), the locomotion was characterized as undirected. But later, the idea of taxis was reserved

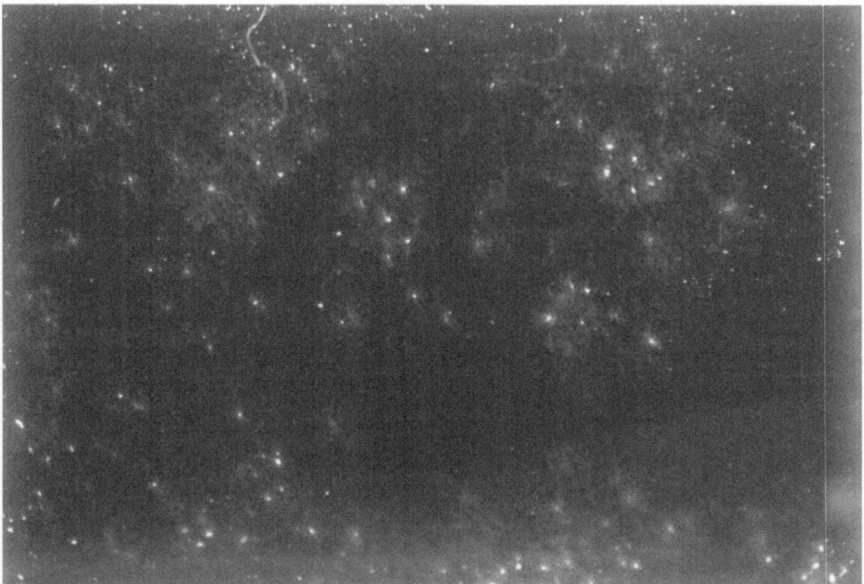

Figure 9.1 A "universe." ***Source*****: Photo by the author.**

for directed movements only; therefore, the paramecium's locomotion was then termed "ortho-kinesis." Since "taxi" originates in the Greek word for "arrange," Fraenkel and Gunn (1961), among other biologists, felt this meaning to be more appropriate to indicate a random turning. Where a stimulus can produce both movement and direction, as in the behavior of Dendrocoelum, a planarian, it is referred to as klino-kinesis. Where some molds that are generally solitary unite to form a stalk and budding structure, this temporary "social" behavior is in response to environmental change that stimulates a fruiting body allowing for some members of the "social unit" to escape the area as spores (Dworkin, 1972). Whether one calls this altruism, which implies recognition of motivation, or simply induced behavior, as in a chemical "taxis," is irrelevant. It is a form of social behavior that the individual Myxobacteria participate in as temporary members of a social unit for a specific end (Wireman and Dworkin, 1975).

Wilson (1975) describes a number of similar types of social behavior, in insects and vertebrates; the essential question is, what is the cognitive life of the individual animal that determines such behavior? Social grouping is a rather late feature of life. Most of the time, that life has existed on this planet has been as single cell organisms, mainly prokaryotic cells (without nuclei) and later as eukaryotic cells (with nuclei). The association of such cells into groups as colonies for temporary or permanent arrangements is the history of the development of multicellularity. Fossil evidence of multicellular organisms is scarce prior to one billion years ago, yet evidence of earlier examples has recently appeared. The joining of cells into colonies and then into single clonal organisms defines the evolution of sex and meiosis.

GEORGE WILLIAMS' ORIGIN OF SEX, COMPLEXITY, THE SUPERORGANISM AND ALLAN WILSON'S COMPLAINT

In the April 11, 1975 edition of *Science*, Allan Wilson published a review of a new book by George Williams on the origin of sex. It actually contained a substantial theoretical work based on a mathematical proof relating to assumptions concerning the evolution of sex from prokaryotic cells to eukaryotic cells and multicellularity, or complex organisms produced by meiosis and the cell fusion that completes the union of gametes. Williams attempted to produce an explanation for the substantial decrease in survival and fitness assumed by most biologists in the process of meiotic reproduction over asexual reproduction.

One can begin the analysis by a simple philosophic point: asexual reproduction allows for a kind of immortality; the cell, by going through mitosis, regenerates itself and, at the same time, duplicates itself! The general idea in Williams' time was that asexual reproduction was the normal process in simple prokaryotic cells. Certainly, prokaryotic cells engage in cell fission as a rule (Margulis, 1981, 1992); mitosis evolved from this condition. One limitation that went along with this idea was that prokaryotic cells would languish in a stage of poor genetic variation, as they required mutations in single individuals to occur to add fitness. Today, we know that bacteria engage in gene transfer via conjugation bridges, a form of "sex" but a means of increasing diversity, especially antibiotic gene transfer, it is probably an early form of crossing over in meiosis. It is analogous to crossing over in meiosis in multicellular cells. But there are many forms of premitotic reproduction or duplication of primitive cells where diversity can appear and be transmitted (Margulis, 1992; Ris, 1975). Genes could be lost and mutate. But prokaryotic cells can also increase diversity by cell fusion. Two cells with different gene sets can fuse, exchange genetic material, regenerate themselves, and then divide to restore individuality (Trevors, 1999).

While the building of eukaryotic cells from prokaryotic cells via processes of digestion and incorporation can be demonstrated from the activities of heterotrophs in either ingestors or decomposers (Margolis, 1992), this is also a means of horizontal gene transfer (this idea is controversial; see Stanhope et al., 2001), either seen as gene transfer (the information) or partial ingestion and incorporation (structure transfer) (Hall et al., 2008). Margulis (1992) argued that multicellularity evolved several times in prokaryotes before the rise of multicellular eukaryotes. Some think it has evolved separately 46 times (Grosberg and Strathmann, 2007). Does this mean that there is more than one track to complexity and to the accumulation of variation than the singular scheme proposed by Williams? These transfers of genes make trees of life difficult (Puigbo et al., 2009).

We can imagine a process where instead of direct cell fusion occurring, an intermediate system arises where the conjugation bridge becomes a pathway for cell fusion controlled within the tube-cell wall matrix as one cell becomes amoeboid and moves through the tube as in Spirogyra and at the end of the journey initiating the fusion process forming a zygote (zygotes are also produced by fission in this group). Here, the production of gametes as external entities, usually producing a lytic release from the "parent," takes on instead the nature of exchange of genes typical of conjugation (a less dangerous and complicated process than cell fusion) but results in an integrated "individual" as when genes are accepted in conjugation and integrated into the recipient's genome. This is a form of isogamy (gametes of equal size) before the evolution of anisogamy (sex gametes of different sizes, as in sperm and ova) (Dusenbury, 2009).

But some organisms like *Ulva lactuca* combine a haploid with mitosis and the production of gametes with a diploid stage, and the joining resulting in zygotes. What is even more complex is the process by which "colonies" of fungal cells create hyphal cross-walls or septa, which allow cytoplasm and organelles including nuclei to move between different hyphal "compartments" transforming the fungal "colony" of individual cells into a great syncytium or giant "cell" (Fleissner et al., 2008). Here, the colony takes on the distinct appearance of a superorganism (not as developed as in an insect society; see Gillooly et al., 2010), as cells of different individuals mingle with their organelles across spaces of a certain "communal" nature. One is led to consider this as the foundation for the development of "self," and the eukaryotic world of cells derived from the clonal expansion of meiotic produced and then fused gametes.

Cells of a *Volvox* colony (a genus of chlorophyte green algae) take on specialized functions and some (especially those of the outer wall) lose their ability to reproduce. Colony specialization certainly mimics what we typically see in multicellular plants and animals. The appearance of such colonies can be demonstrated at about 2.1 billion years ago with the development of oxygenated environments (El Albani et al., 2010).

Given that meiosis is an evolved and more complex form of mitosis (Wilkins and Holliday, 2009), it is relevant to investigate how sex has evolved in the context of increasing complexity from the pre-Cambrian to the appearance of multicellularity about 600 million years ago. If we agree with Margulis (1992) that there has been considerable and continuous transfer of genes and organelles (like mitochondria) that have originated in the incorporation of simple organisms into more complex organisms by partial digestion, then the process of addition and integration has been an avenue to complexity. Yet, how do we understand the evolution of self from colonies? How do meiosis and cloning produce the self versus the other?

Research with filamentous fungus like *Neurospora crassa* has demonstrated how colonies of these organisms grow up with cross-walls or septa, which allow for cytoplasm and organelles, including nuclei of different cells to move between different compartments of the colony producing what can be regarded as a supercell or syncytium. This intercolony transposition of cellular components gives the impression of a unit of individuals that could be seen as a giant "self." In multicellular organisms like ourselves that have developed from gametes produced from meiosis, our cells remain individual units but regard and communicate as one individual. This is the great achievement of the Cambrian, but while this process of multicellularity invented sex through the joining of gametes' nuclei into integrated pronuclei of the fertilized cell, it also created a temporal being with a distinct life span. Sex, therefore, trumps the eternal life that bacterial or fungal organisms, who engage only in fission, fusion, mitosis, or conjugation, enjoy. Thus, complex life is both rare, in that it appeared late in the evolution of life on the planet, and short-lived. Complex animal societies are also rare by this measure, but more so, yet also rare in their evolution in the vast numbers of animal species that lack complex social life. They are also short-lived in that they require frequent, if periodic, renewal. This pattern does seem to fit human societies as well.

Human societies have been seen as superorganisms, yet these units, like other animal societies including ants, bees, wasps and termites require renewal, either with new queens or new locations where the hive or colony moves and often also produces a fertility event for the queen or new queen. The problem of human society is that renewal often has involved war, civil war, disease, or migration. One wonders if the future of human society as globalism accelerates and becomes more integrated will produce integrated information cities as Sassen (2012) has envisioned where a global human superorganism will become stable and achieve renewal without destructive means. While relocation is theoretically possible, its achievement in time has been doubted (Sagan, 1966).

It is often argued that humans have produced the only complex societies, domestication of plants and animals, structures, language, and art. This is an ignorant argument as numerous animals have achieved these features of social life, thought what is art is so opaque that few venture to define it. E.O. Wilson has been one of the foremost students of complex social life in animals and their products. Various species of ants and termites have species of fungus they "farm," including water, weed, and fertilize. Structures built by these insects are complex and often provide for cooling in summer and retention of heat in winter. Bonner (1980) summarized the evolution of culture in a variety of animal species with complex social life. We can find parallels among other social animals as I have suggested in part I of this book. He finds most telling the inventions of the macaque Imo on Kashima Island in the 1950s (mentioned previously) whose invention of washing sweet potatoes in saltwater spread to most members of the troop. Here, one individual could create an adaptive change in society by her perceptions. To go further has been the task of some sociobiologists like Wilson (1975) and Bonner (1980). Yet, on a more detailed plane in human complex societies, scientists like Amitai Etzioni (1988, 1961) have dissected human societies and assessed their institutions to determine if there are qualities to human societies that reflect ideas of human behavior like "rationality."

Etzioni (1988) finds that institutions, including corporations, firms, government agencies, permeated with irrational behavior and managers who engage in the use of "rules of thumb" that seldom accomplish the goals for which they were enacted. Here, however, we need to distinguish between ideas of rationality, logic, and success. A pigeon in Skinner's experiment may feel positive when the reward is delivered given its response of "superstition." Humans act within a complex system of conditioning, defined by symbols and both subtle and objective signals, that we call culture, and since we grow up in the linguistic context of this signaling that structures our cosmology, we seldom question the "superstition" that guides the rationality/irrationality of our behavior—the stock market being a good example of institutionalized irrationality (Caldararo, 2009b, 2011).

Much is made of predictions of the stock market and yet success seems most elusive, defeating some of the great economists of our time including Keynes and Schumpeter. Humans have attempted to form knowledge of the future by myriad means, including studying the entrails of animals, yet it is both the desire of the individual and the interpretation of signs that defeat such attempts.

Modern global culture is dominated by the ideas of progressivism and modernity (really a consequence of the ideology of progressivism, see Almond et al., 1987). In this ideology, there is a constant expectation that society is improving and increasing in perfection, which is often defined by technological development. Modernity argues that this situation has never existed before, that both the increase (and desire to create innovations) is a part of progressive development and that this process is only a part of present complex society and its institutions. On the issue of scale and progress, Geoffrey West (2017), extending the work of Lotka, has remarked on how scale can be seen as a fundamental means of understanding the functioning of systems and sees, much as did Herbert Spencer or Plato, a universality of laws behind complexity at all levels of organization. However, he is curious about why economists have been blinded by progress and see innovation and development as panaceas, rather than recognizing limits, for example, entropy in systems as complexity increases. Like Lewis Munford and Jane Jacobs, who promoted the idea of the adapting city as a self-organizing entity, the emphasis has been on a progressivistic view of complexity rather than seeking concepts of sustainability.

But it is necessary to separate these ideas of progressivism and complexity from the hubris of rationality and purpose in social life. Almost every society we have studied has believed that it was the best that ever has been, with some few exceptions. The idea is always the same, be it today or among the Victorians, the Romans, or Confucius' China: idealization of the golden ages past. In some cosmologies, in India, for example, the loss of the golden age, like that of the loss of Eden, is due to human failure. But just as the idea of Eden receded into an exotic distant land as human technology increased the complexity and density of social life, the nature of the golden age became obscure. While humans could be seen as less perfect than those of the golden age, their qualities and that golden life seemed less possible and vague (Campbell, 1949).

Yet, even as the idea of Eden recedes from earth to a mythological location, the behavior of the golden age also becomes more obscure as to what people should be doing and why they exist. A contradiction develops between what is possible for people to do in life that is expected and the given of the golden age. Here originate religious explanations that provide new avenues to that place with the divine, and in this, we find a growing separation in how complex society sees the world. Most traditional, simple societies have gods that are nature or represent nature; humans are part of that nature. Complex societies, especially as they increase energy extracted from the natural world, either in food production or resource extraction, we find humans have created gods that look and act like he/she does. We see this retained, perhaps in India's Hinduism as "nature as gods," though there are many forms of this religion (Thapar, 2002). Humans are increasingly seen as outside of nature and nature is inert and a source of human ambition, but no longer divine.

One might say, in opposition to the earlier views of primitive/simple versus complex/civilized human consciousness, that instead of primitive man being less rational, he seems more so. Though Malinowski (1916, 1922) considered "primitive" man to be as rational as "civilized" man, and the opposing view is usually attributed to Levy-Bruhl (1923, though it originated in much earlier writers and travelers), in comparison with "modern civilized man" we might consider modern man more prone to irrationality. Therefore, it is difficult to utilize these terms to differentiate the behavior of humans and nonhuman animals. Irrationality does seem to be a central type of behavior, organized and embedded in complex society.

We call on nonrational behavior as a means of motivating. This can take a multitude of form, as in armies' murderous acts (e.g., Romans, Mongols, or modern mechanized versions like British in WWII, current "drone warfare"), Christian pogroms of Jews, the execution of "enemies" of the state in the French Revolution, massacres in Rwanda, the colonial terror against native peoples of the world by Europeans, or Mao's Red Guards of China's own people. Such mass behavior recruiting and control can be seen as related to the nature of complexity, whether in humans, ants, or bees. Similar behavior, revolutions of less dramatic mass movements can be adaptive to change institutions and be characterized by more logical and rational communication (e.g., the Renaissance) yet such changes are difficult to assess in the time frame we have available. Assessing the results of human complexity has been the focus of environmental movements, yet these have lacked any clear definitions of what human society should look like to be sustainable or even if human society can be sustainable (Caldararo, 2013). This has often been associated with religious ideas, the earth as a gift from the god is often related, or as a garden to tend and protect or as a being to cooperate with (Gaia principle).

There has been too short a period since the transition from hunting and gathering to food production in the Neolithic to determine if sedentary behavior and complex, dense social life is adaptive for primates. War and disease may continue to play roles of dispersal (keeping populations separate) or to drive innovation. Resource depletion, especially water, has been part of this dispersal process in the past (Redman, 1999).

For example, historical data on California's water shortage (which applies to much of the western United States) seem to show that there is no solution, especially in how the data define the damage to groundwater supplies and projected rainfall, given also projections on population growth and water needs. This is not new, though in the recent past, alternative sources including recycling water and desalination were important, but now desalination is considered by many environmentalists to be unacceptable and to only promote development. Ester Boserup (1965) demonstrated how increasing population in Mesopotamia during the Uruk period led to efforts to irrigate more marginal lands, resulting in falling yields and disastrous salinization. Marginal returns and reduced returns from degradation due to this intensification of agriculture produced a downward spiral. Gibson (1974) detailed these effects across the region at the same period. This degradation took place in the process of increased population and urbanization where small villages and towns were abandoned for larger urban centers, and the pressure of these concentrated populations and their increased needs created instability and recurrent crises, according to Robert McC. Adams (1981). This same trend of increased urbanization and decreasing rural populations has been taking place again in a global scale since 1900. Saskia Sassen has been describing this process in relative terms beginning with her 1991 book, *The Global City* as well as describing the negative aspects in later texts (e.g., *Elements for a Sociology of Globalization* [2007] and *Cities in a World Economy* [2011]).

Her recent book, *Expulsions: Brutality and Complexity in the Global Economy* (2014), contradicts the rosy predictions of urban futurists. An ever rising population, demanding a middle class life style and living in dense conditions was not a successful adaptation in the past and is unlikely to be so in the future. The hubris of modernity has created the belief that we can continue to expand our population and resource needs without end. But ends are the normal conclusion of expansions. Fear of the reduction of fertility in Japan is irrational, and largely arises from westerners (e.g., Lewis, 2014). Yet, such arguments are often linked to criticism of Japan's reluctance to open itself to the kind of massive immigration typical of America and the West in general. Yet, if demographers' projections for a leveling of human population growth by the mid twenty-first century are correct (see Mueller, 2001 for an alternative view), Japan can be seen as in the forefront of that change. Japan's reduction is similar to periods of adaptation to crises in the past and may provide a model for population stabilization for the rest of the world (Caldararo, 2003b). Yet, that stabilization may be far off (Sullivan, 2013). In fact, one has to ask: why do we need an ever-increasing population? It seems the only real answer is to maintain consumer society (more buyers), and yet that system is based on the origins of the state where an ever-increasing population gave rise to more symbols of power and prestige, for example, monumental building, grand armies, and food surpluses. (These are also essentially the goals of Bronze Age societies.) It all seems clear proof that, with the huge brain, we have still failed to solve our most pressing and threatening problems: population, war, and disease, which undermines Neel's (1970) theory that human ability to regulate population was one means of driving our evolution. Perhaps our failure to do so, to restrict or eliminate the means (abortion, infanticide, etc.), is evidence of the immaturity of our eusociality or its form as a maladaptation in Lowie's (1917) sense, which may prove fatal.

Disasters, defeats, disease where a people suffer reversals of fortune, and be driven from the land of their god, taken in captivity or dispersed by a conqueror can cause new definitions of god or cosmology. These can bring ideas of the basic cosmology into focus and question assumptions of the relation of the people to the god. This is the role of the "religious formulator" of Paul Radin (1937). In post-Republican Rome ideas that the gods had weakened, died, or abandoned the earth due to the lack of potency of the prophesies was one kind of response; another is that the gods were fickle, jealous, mad, or irrational beyond human understanding. This has often been the means of reaching an adjustment in the Judeo-Christian-Islamic tradition. But, failure and defeat do not always result in redefinition or reassessment, basic elements of cosmic relationships can persist unchanged even in the face of thousands of years of setbacks as in the case of Judaism, especially after the triumph of Christianity. But, as they are brother sects, both claim special relationships with the central god of the tradition who has promised ever-increasing benefits to humanity (Genesis 9:7, "be fruitful and multiply," for example). Such irrational, or maladaptive in Lowie's (1937) definition, attachment to elements of religion and magical thought (Vyse, 1997) are serious impediments to redefinition and change (Caldararo, 2002c). As the world's religions are divided into two main groups, the Judeo-Christian-Islamic tradition and the Hindu-Buddhist tradition, as groups that have historically expanded via conversion and missionary work with an ideology of increasing numbers of members, competition for members intensifies other economic and political conflicts (Andrews, 2010; Arnold, 1913). This is less true for Hinduism until very recently (Sharma, 2011); however, Buddhism evolved from it and has had such a conversion value, yet not free from conflict (Reischauer and Fairbank, 1958).

Yet, if the god of the Judeo-Christian-Islamic tradition has failed to protect and give power to the origin group—Jews—and yet they have not abandoned him, power has been transferred to the Christians and then (after the eighth century CE) Islam, then back to the Christians after the fifteenth century. How do we assess the resistance to adaptation to the problems of mass society and overpopulation, for example, poverty, war, inequality, environment destruction? As Communism threatened capitalism (based on Christianity as an economic system, though Socialism/Communism in its origins as a theory had made similar claims) in the twentieth century, the response of the capitalist west was a rejection of its own ideological failure and a renewed struggle against Communism. While the originator of Christianity taught poverty and love, the evolution of the Catholic Church and the Protestant denominations that fractured from it, redefined that message into one of conquest and forced conversion (e.g., the history of the Americas). Logic in systems of belief is not a constant feature, and new definitions also explain defeats or setbacks, as in ideas that Protestants were led by the devil, or similarly for Muslims or Nazis, or that all bad things come from god's enemy, Lucifer. Alternatives to belief systems can be threatening, as in communism versus capitalism, Judaism versus Christianity or Islam. What has been true in the past 10,000 years is that complex societies have expanded at the expense of simple ones. Simple societies have been incorporated, conquered, and driven out or subjugated and their land seized. Tolerating the simple society alternative has been impossible, a relentless extermination has taken place instead, eliminating the possibility of a retreat into the cultural complex of simple societies rather than a collapse into forms of feudalism, which are supported in part by the surviving technology of complex society without its institutions. This wave of global destruction of simpler, traditional societies continues today across India, through Indonesia, the Americas, to Africa, and back to places like Waziristan in western Pakistan.

As Nietzsche (1888), and later Harnack (1902) argued, the message, combined with conquest and empire, became the opposite of the loving Christ and community of the Gospels, and was transformed from a message of love to its opposite. Just as Weber (1948) searched for the differences between Protestant and Catholic in economic development, McClelland (1961) searched history and across cultures to find the key to motivations behind achievement no matter how defined. The hydraulic theory of oriental despotism defined by Wittfogel (1957) described a system of government based on a cultural foundation for the East (including China and India) that limited individual achievement and innovation. This idea was undermined by the voluminous work of Joseph Needham (1981), who demonstrated the vast innovations and scientific developments of traditional China. Also, a number of archaeologists took issue with the scheme, including Robert McC. Adams (1966). While Karl Marx and Arnold Toynbee stressed the effects of economic and ideological struggle in society as an important motivation for change, the idea that class struggle over resources, wealth, and prestige is the motivation for innovation is an apparent but not comprehensive explanation. As Needham (1981) shows, capitalism and class struggle are not necessary for technological and scientific innovation and discovery. In fact, many of the innovations that gave power to the west in its colonial conquests, like gunpowder, were derived from Chinese inventors.

We are in the midst of a new period of modernity, and every society that achieves hegemony believes it has invented modernity whether it is Ancient Egypt, Han China, or industrial England (Caldararo, 2013). Power claims the field and the apogee of civilization. Jaspers (1953) notes the difficulty in establishing claims of modernity. Yet, Landes (1999) argues, as many before him, that China restricted freedom and, therefore, progress. This depends on what one calls freedom; for example, Polanyi (1944) describes how thousands of English farmers were driven off their land and made homeless and penniless by the Enclosure Acts, and other acts that followed like the Poor Laws restricted wages and movement of workers, creating a virtual police state requiring registration for work (Webb and Webb, 1927). Thus, one cannot say that class war creates innovation any more than freedom, slavery, lawlessness, or chaos can achieve. The outcome of complexity and inequality together, in each case in history, leads to "globalism" to counter conflict (disorder-entropy) at home, and whether it be in Athens, Rome, the British Empire, or the current economic system, the struggle for cheap labor and resources drives empire and leads to greater forms of entropy with different trajectories for collapse. Rostovtzeff (1926) describes this process for Rome, and Tainter (1988) draws attention to numerous factors involved and the complications that can result.

So, for many Americans, the stock market goes up and down with the action of unknowable forces, just as climate change seems a hoax perpetrated by evil scientists who work for a godless government. Those who argue that climate change has happened in the past, for example, in the Ice Ages of the Pleistocene, do not understand the current danger of a runaway greenhouse effect (Kasting, 1991; Smil, 2003): once started, it might not only make life unpleasant, but could achieve a Venus-like hell. If logic and reason are learned, are not illogic and irrationality also learned (Boas, 1911)? But there is another possibility, one that requires not devils, but chance and accident. The polytheistic Romans and Greeks conceived that the gods were like men, and so one could not expect consistent or rational behavior from them. For other peoples like Melanesians, the driving force in the world is an impersonal force, Mana, which can be controlled to some extent but seems uncontrollable in many contexts. If the universe is materialistic and there is no divine force, then this idea is surely close to a scientific one—but it can also be seen as both rational and irrational (Morris, 1987). Are humans then the pawns of accident, and is the human mind a blank slate floating in a sea of coincidence that it must learn to see as rationality or fated reactions of accidental molecules?

MIND AND SOCIETY

What we have been concerned with in the first part of this book is not whether humans were alone in the production of culture, since we know that other animals have achieved forms of behavior that are both protoculture and are elements of complexity (Bonner, 1980; Griffin, 1976), but our question is why it took so long to develop human culture and complex human society. Related to this is a search for what features of human physiology, especially the biology of the brain, might have been crucial in this process.

We touched on questions of mind and the nature of consciousness, yet associated with these ideas is a question of the place of humans in the universe and the nature of reality. This is a topic that is relevant here as it relates to the possibility that human self-knowledge is a selective advantage or that it is a disease of human nature, producing psychological pain and social pathology (Laing, 1959). To some extent, this question impinges on ideas of free will and debates in the scientific community over sociobiology (Caplan, 1978). Gowdy and Krall (2016) broach the opposite question concerning the evolution of the superorganism from eusocial foundations. It is possible that the evolution of the human mind, under social selection, is already closing, partly as a consequence of self-domestication. A part of this process is likely to be the evolution of a capacity for belief, whether religious or ideological, that so conditions the individual mind that it closes off the potential to revolt, resist, and remodel hierarchy as heterarchy. I have already discussed the process of mass psychogenic disease in the context of fundamentalism (Caldararo, 2002c, 2004c), which touches on this closing of the true believer. But to the question of what opposition to the current economic "principles" that Gowdy and Krall (2016) argue drives the destruction of the environment and evolution of the superorganism, we do see leveling attempts (as in Occupy Wall Street) and millenarian and revitalization movements (e.g., the Naxalites in India), which can end in the renovation of existing social institutions. Some appear as more oriented to destruction of the dominant order, as in the Khmer Rouge or ISIS/ISIL (Caldararo, 2015b), or the political philosophies as in the Russian Nihilists or European anarchists (Caldararo, 2004c). On the other hand, such organizations seem to cause a decrease in freedom and liberty in nations under assault and might have the opposite effect of accelerating superorganism development. Wilson (2012) and Armstrong (1989) emphasize aggression against outsiders, but human history seems as filled with rebellions, intrigue, assassinations, and intragroup conflict and violence (e.g., in Rome, see Appian, 165 AD; or for the city of Florence, see Machiavelli, 1520–1525). In terms of selection, however, for Wilson (2012) and Armstrong (1989), to put an emphasis on the value of conflict in promoting human evolution, there is a significant problem. If genes are related to success, then how do we consider the fact that events often do not reward those who are the inventors or creators? For example, the Spanish defeated the Inca with innovations (and disease they had acquired) they had not invented, both in terms of their ships, cannon, guns, and strategies (Cipolla, 1965) as well as domestication of the horse?

WHAT WE KNOW AND HOW WE KNOW IT

Discussions on the theory of Sociobiology usually begin with its author, E.O. Wilson. In this case, however, since I am an anthropologist, I will begin with an inquiry into the nature of knowledge. There is certain hubris in this idea. Wilson conceived it in his introductory text, not as a theory but as a law of biology. This attitude has suffused many researchers in biology who have carried on his work. Consider, for instance, Richard Dawkins in the opening lines of *The Selfish Gene* (1989 edition). He states, “Intelligent life on a planet comes of age when it first works out the reason for its own existence.” This is akin to the Victorian idea that savage people could be distinguished from the civilized by the fact that they did not know their parents. Of course, they did know them, but not in the way the missionaries believed heathens could and so did not even consider their different kinship systems as valid. The failure to examine the nature of an object, animal or universe, is based on an arrogance that knowledge is confined to that which one knows, that is, what one is familiar with as knowledge. Griffin (1976) argues that this problem has confined studies of animal awareness and language for the past 2000 years.

In Dawkins’ case, as in that of the Victorian, this is a most curious belief. We know now from over a hundred years of anthropological research that non-Western peoples did know their parentage; it was racism and ignorance that prevented Victorians of all nationalities from seeing the truth (Montagu, 1964). Ideas of parentage are so embedded in biological theory that human family history is constructed on Victorian ideas of monogamy, when we know that, for example, Andaman Islanders (Ratcliffe-Brown, 1922), Medieval Icelanders (Magnusson and Palsson, 1969), and the Inuit (Weyer, 1932) practice forms of marriage and child-rearing that are distinct from modern Western ideals. For the Inuit, wife sharing and swapping would certainly upset ideas of inheritance and self-interest in raising children so championed as the foundation for the evolution of the human family, society, and brain (Deacon, 1997). The Andaman Islanders give their children to other groups as a means of maintaining good relations. Thus, raising your own children and ideas of self-interest and inheritance are upended from Western conceptions (Radcliffe-Brown, 1922). Essentially, many who write on the evolution of gender relations in humans utilize either contemporary idealized conceptions or philosophical ideas like those of Trivers (1972) and Dawkins (1989). Human relations are more diverse, both now and in the ethnohistorical record, and their models distort our understanding of gender roles, child-rearing, and inheritance patterns.

In the case of Dawkins, however, it is a broader ethnocentrism that is at work. Or is it anthropocentricism? One like that of Alexander (1989), who argues in like manner that all humans know what is best for them, like all genes and their gene machines? Dawkins goes on to clarify, "Living organisms had existed on earth, without ever knowing why, for over three thousand million years before the truth finally dawned on one of them." He means Darwin, of course, and nowhere in his text is Alfred Wallace, the codiscoverer of evolution, mentioned. But then, this is also consistent with Dawkins' argument that, to be a good Darwinian, all things must have one beginning. This is also a primitive idea since almost all origin myths begin in the same way, and our Western religious system is based on the god who will have no partners in creation.

But let us consider Dawkins' claim here. No other organisms on earth in its long history of life have ever known why they existed. How can Dawkins know this? Has he some crystal ball to travel in time to interview long dead plants and animals? Has he found the fabled King Solomon's Ring that allows him to speak to all living organisms? Is he a modern Odin whose two crows, Huginn (thought) and Muninn (memory), have traversed the universe for him to exhaust the possibilities of self-reference in animals? Of course not. But Dawkins is doing something remarkable in this and his book *The Extended Phenotype*. He is defining the relationship between origins and ontogeny. Often when evolution is discussed, it is taken for granted that organisms evolve from fertilized cells to become adults and that this process is inconsequential to us except that it takes place on a regular basis providing an environment with animals and plants that look and function enough alike to be recognized as similar. The fact that Dawkins focuses on is that Darwin was aware that this seeming similarity hides a vast amount of variation when one looks closer and begins to measure individuals and study how similar they are to each other. What Darwin found is that this vast variation is the basis on which Natural Selection acts to produce evolution. Not only is his book *Origin of Species* a discussion of variation, but it is also in many ways a digression on his study of the variations produced by domestication of wild types in his book *The Variation of Animals and Plants under Domestication.*

But Dawkins goes further and proposes that it is not the individual we should be concerned with, but the gene. We must recognize that this seems absurd at first. Individuals appear to us as actors on the stage of life, how can we ignore them and concentrate instead on simple sequences of base pairs in DNA? This feeling of absurdity turns out to be a good one, for while Dawkins' idea is compelling, it leads us away from how evolution works to the mechanics of evolution. It is easy to see how he has fallen into this trap. Dawkins tells us at the beginning of *The Selfish Gene* that other authors have gotten it wrong in attempts to understand how evolution works. He claims to have discovered its secret. But he arrives at his answer only by ignoring how organisms work and how genes are based on generation after generation.

Dawkins is presenting an atomistic view. He is seeing all the genes that make up an animal as being not only independent in their struggle for existence (being passed on), but as being separate from their function in the body, the cell, the genome. Here, the intron is at war with the exon, the sequence of base pairs composing the message for an enzyme, in struggle with that for its regulation and all other proteins. The central goal, and the only one worth considering in his view, is the outcome of the successful transmission of the "gene" to another replicator, another body. This would certainly be the anthem of a virus, or a viral gene. But we find in biology that the nature of life is not only wholly related to competition, but rather, cooperation. The genes function within the colony of cells that make up a human or other animal. On the other hand, if we accept Dawkins' idea, we cannot assume that the process of competition is very straight forward, since as Doolittle (1999) has shown, and Crisp et al. (2015) support, gene sequences have been transferred across genera, families, and kingdoms over the past half billion years or more. Thus, the same gene sequences have been at war against themselves. This is certainly an interesting thought, but one that contradicts the general thrust of Dawkins argument. Gene sequences have then used their replicators not only to spread their specific sequences but also to compete with themselves in different replicator bodies. More complicating is the fact that most of the genome is "silent" in many organisms, believed to be nonfunctional, and not just in reference to introns, which now seem to have regulatory roles, but in terms of endogenous retroviral sequences, which are essentially along for the ride with no observable role whatsoever.

But when Dawkins attempts to explain why independent genes joined together to form cells, he argues that by "ganging up together," their biochemical effects might have complemented each other (Dawkins, 1982). Using the same logic one could say, I suppose, that linked genes were also "ganging up." As Samir Okasha (2003) notes, this is cooperation and thus an argument for group selection. Here E.O. Wilson (Wilson and Hölldobler, 2005) has modified his position on kin-selection versus group selection (to a certain extent thereby vindicating Lin and Michener's 1972 theory of the evolution of sociality in insects via mutualism and defense). Bonner (1980) emphases the benefits from association and complexity, whether temporary as in some fungi or permanent as in the clonal cells of our bodies produced via meiosis and sex. Is the functioning of the entity produced by a mass of amoeba (like *Dictyostelium discoideum*)—the "bodies" of cooperative sponge cells that can be disaggregated yet reassociated to recreate the "body" and the cells of our bodies—the essence of "mind" that is so often referred in a circular fashion as "self"? Perhaps this is too metaphysical, and the interactions of these cells, as in the synapses of our brains, are just mechanical, chemical, and electrical connections that briefly (relatively speaking) produce the illusion of "life" or "mind?"

If we relate this approach to Steven Pinker in his book, *The Blank Slate* (2002), we find he characterizes the current scientific view on nature versus nurture in very absolute terms. One would believe from reading this book that most scientists are in agreement with Pinker, or that the only dissidents are just anthropologists who refuse to face facts. This is not the case. As David Hull puts it in his review of Pinker in *Nature* (Hull, 2002), Pinker is beating a dead horse with a straw man. But Ludvig (2003) shows that Pinker exaggerates and even misquotes and misrepresents the positions of Watson and Skinner. The debate Pinker argues versus environmentalist positions never existed; even Locke realized there were "innate capacities" that mediated human development. Some of my comments here benefited from conversations with Ian Pitchford, editor of *Human Nature Review,* in 2003. I am responsible for all of the content, however.

Pinker is selective in his use of studies and ignores serious problems with the data in most. This is not unusual, as Pinker has applied the same approach to his study of violence (Pinker, 2011). Here, he has used archaeological and ethnographic data and compared it to that collected in different societies at different times over the past 2,000 years. This is not in itself a condemnable method, but he ignores the different contexts: for example, hunter-gatherers may be more or less violent depending on who is recording the information and what their attitude is toward the hunter-gatherers and theirs toward the recorders. Another consideration is the long history of humans as hunter-gatherers versus sedentary human social environments. Sedentary societies are able to produce more men, train them for violence, and engage then in organized aggression.

We do regard war differently in different societies and philosophical contexts (Caldararo, 2004c), but armies are likely only 5,000 years old, mechanized war only some 2,000 years old (e.g., Roman war machines), and the machines of mass murder only some 200 years old, with doomsday weapons only some 80 years old. A considerable portion of the literature about war collected in the colonial period has been questioned, as it ignores the effect of colonial armies, policies, and the population pressure of colonists on native peoples around the globe (Ferguson, 1992). These would produce effects of increasing conflict and war numbers in this earlier period. Ferguson (1992) argues from evidence of the Yanomani that violence increased on contact, partly due to competition for western goods, especially steel axes. In a strange coincidence, Johnson, Salvatore and Spierenburg (2012) have produced data from a wide variety of methods and sources indicating that during the early colonial period from 1450 to 1750 CE, there was the most significant drop in European violence. An explanation for this might be that the migration of European population to the colonies focused violence abroad; yet, this argument is questionable, given that the Wars of the Reformation took place in this period, as did the Inquisition, as well as the Turkish invasions of Europe. It is interesting that they find a correlation between the reduction of social rewards for male fighting (armies, duels, etc.) and the rise of aggression involving women becoming more visible. Also, honor killings, and an increase in violence and murder that took place among the working class also tookplace with a general reduction of the same among the upper class (Spierenburg, 2008).

Nevertheless, war has led to more efficient weapons, and the massive battles of the first complex societies to today's regional and world wars where vast areas and peoples are involved. Complex societies are only about 7,000 years old as a human phenomenon, yet the invention of war, and massive war is a consequence of these complex systems of humans. Pinker (2011) seems to ignore the fact that the systematic execution of war and war innovation has been increasing both in scope, extent, and the effects it has on civilians.

MADNESS, ABILITY, AND DOMESTICATION

Wilson (1988) among others has argued that our self-domestication has had a significant effect on our overall behavior both within societies and between them. Elias (1978) has placed emphasis on the process of standardized changes over time in history that have modified acceptable behavior and created more pacific populations, at least in times of peace. A cost or consequence of this modification and domestication may be madness and psychological distress especially the depression we see so widespread today

(Scull, 2015). The idea of mental illness differs in different cultures (Wallace, 1972), but does mental illness have a role in human social adaptation or our evolution? Some social movements and mass psychogenic behaviors appear irrational by some standards yet can appear adaptive in their circumstances (Dalton, 2000). The concept of "thinking" and madness or mental illness is tinged with cultural hubris and racism. What is one person's "idea" is another's illogical perception based on cultural contrast (Levy-Bruhl, 1923; Murray, 2007). Is mental illness a response to stress, a social indicator of societal maladaptation, a result of creativity, or a means for new social perceptions that lead to social evolution? The debate on this topic is considerable and ongoing (Woolfolk, 1999). But the complexity of the interactions of genes in the brain often lead to disturbances that can be considered illness as in schizophrenia where synaptic pruning, so important to the development of the human brain, is involved (Sekar et al., 2016).

Maish (2008) described the avalanche of stimulation people feel today in the connected world as a negative pressure. Huizinga (1954) opens his book on the transition from the Middle Ages with a description of the noise of medieval towns and the public displays of prestige and cruelty. Ekirch (2005) discusses the addition of lighting to towns in the advent of current modernity, though towns often had forms of lighting in various complex societies of the past.

While our way of living has changed dramatically since the beginning of the Neolithic and the end of nomadism for most humans, the human millennium has conquered man (Wilson, 2012). Henry (1963) argued that this change brought on suffering and required mass psychological changes that caused madness both in crowds and the individual. What is normal, however, depends on the social values. Yet, human behavior over the past 10,000 years and certainly the past 2,000 years is rather abnormal for the life of a primate.

Complex societies require some degree of subordination for recruitment and order, we note that wolves can organize a hunt around the dominant animal and others defer in gesture and action. The posture of the subordinate, kneeling, bowing is what we see in religions of complex societies today where the individual takes on the posture of the suppliant, in submission like the defeated or prey to the predator. The dominant here is an abstraction, often inculcated from birth (with the few exceptions of converts) and sometimes publically displayed in mass exhibits of prayer.

One central example to Pinker's thesis should suffice, that of twin studies. Pinker seizes the idea of twins, mainly monozygotic twins, to pronounce victory in the genetic side of the argument of nature/nurture. Genes seem to create mind, the talents, attitudes, personalities, and preferences (among other traits) are so similar as claimed by many psychologists who have written on twins, that learning, or the environment can have little role (e.g., Mackintosh,

1998). But, is mind simply a complex arrangement of conditioning, organized in a person's lifetime and expressed or mediated by learning language as a mechanism for efficient responses in words as releasers of coincidence of information? Does this not better define the variations in learning and failures in "understanding" both in situations and conversations? This would seem to fit into Crick's (1995) scenario mentioned above and perhaps that of Edelman (1987).

The area of twin studies concerning research in IQ has a checkered history. While I briefly referred to IQ studies in part I, I will return here to the issue again. One good example is Sir Cyril Burt, who was guilty of not only fabricating his data on identical twins, kinship relations in IQ and declining levels of intelligence in Britain, but claimed the achievements of others, namely Charles Spearman, the inventor of factor analysis. Steven J. Gould detailed other problems with twin studies and studies of IQ in several books and articles, most notably in *The Mismeasure of Man* (1996). To read *The Blank Slate*, one would think that Pinker reflects a general agreement on inherited ability based on later twin studies, but this is also not supported by the literature. Researchers like Plomin (1994) and Bouchard (1999; with Markon et al., 2002) consider genetic components to be more influential than the environmental view of the 1950s, but the problems with research in IQ has been demonstrated by Kamin and Brock and Dworkin (1976). A recent report by Kamin and Goldberger (2002) listed in detail the deficiencies common to all the studies in twins and IQ research in general. The basic design of most of these studies is tainted by a number of organizational problems and problems in analysis and reporting. While an understanding of why individuals differ in ability is of interest, the strident claims for "natural abilities" that are inherited are unsettling and usually proven in time to be false—like the claims of noble Roman families to great natural ability generation after generation, which is proven to be due to the adoption of children of exceptional talent, the desire to establish some lineage which is special in and of itself has been the pursuit of many and the basis of myth and tragedy (Lindsay, 2009). Examples are the royalty of Europe whose inbreeding led to widespread hemophilia and idiocy along with general homozygosity. What we are often left with in the end is that talent can be found issuing from the most humble circumstances, and merit is had not by breeding but by hard work.

In the second section of Pinker's (1994) book, he notes that genes cannot account for all that we see, and both the genome and environment fail to explain the complexity and variation we see. This is certainly true: it is the shuffling of genetic material in the grand potential of a diverse population that gives rise to the surprises we identify as genius and ability. Nevertheless, as I have stated in an article in *Human Nature*

Review (2002), our understanding of the details of how genes function is unfolding before us now. Here, the effect of methylation of histones and histone modification's in epigenetics are an example, especially where environmental influence is demonstrated (Fraga et al., 2005; Yuan and Zhu, 2012). The simplistic ideas of single genes controlling behavior is being proven false and reductionist, like "just so" stories. A good example is with the idea that one gene controlled human brain size and shape (Jackson et al., 2002). The authors believed their work demonstrated that microcephalin, a mutated gene, was the cause. In another study, they believed a gene called "asp" was responsible (Bond et al., 2002). But when another group used a mouse model to create features of size and shape in mice that appeared to mimic human brains, a third mutated gene was suspected (Chenn and Walsh, 2002). Recent research indicates a more complicated situation where the novel microcephalin mutation can produce different phenotypes, one with retardation and one with normal intelligence (Ghafouri-Fard et al., 2015).

Pinker's view is one where IQ is like a great bowl eternally filled with ability. What we see in the evolution of hominid brains is a status between the appearance of the first hominids, like Sahelanthropus tchadensis at about 7 mya to about 2.6 mya, with the appearance of the first stone tools, at the level comparable to the brain-to-body ratio of contemporary apes and a "take off" in brain size growth thereafter. It is only after the appearance of early *Homo* that the hominid brain begins to increase in size and go through the changes, which are associated with cognitive abilities. However, there is a slow increase in brain capacity through *H. erectus* up to the lower level of contemporary humans (ca. 900 cc) to about 600,000 BP. Then, there is another long period of increase to about 100,000 B.P. where the greatest increase is seen in the Neanderthals, as a population, a group which is not even seen as on the human line any longer by many scientists (Caldararo and Gabow, 2000; Caldararo and Guthrie, 2012; Stringer and Andrews, 1988). Thus, if brain size alone was important, why did the Neanderthals become extinct in competition with the smaller brain sized anatomically modern humans (AMH) (as a population)? In my opinion, the Neanderthals are on the human line of evolution and did not become extinct (Caldararo, 2002b, 2016b), but as Deacon (1997) put it, Neanderthals were likely small populations and swamped by the arrival of AMH, but as in Native Americans, they were simply genetically overwhelmed, though we now know they also interbred with the AMH. Nevertheless, how do we explain the decrease in brain size between the Upper Paleolithic and the present? One might refer to concepts in mathematics and systems studies in biology where an increase in complexity produces significant costs; eventually, as Weiner has noted (1948, 1954) after Gibbs, there is an increase of entropy. Quinton (1978) has allowed that some degree of selective advance to the foundations of cooperative social life is to be expected. This is the tentative conclusion we came to in part I.

Still, the idea Pinker is pursuing is tied to nineteenth-century positivism and the idea of human perfection. Certainly, there are other aspects of complexity we can point to in physics and biology: for example, smaller elements are more stable; as we go up the periodic table to the rare earths, we find increasing complexity in terms of protons and neutrons and number of electrons, but these are increasingly unstable. Instability in complex social systems seems to be controlled by the means to distract the population (games, work, war, etc.) and engage a daily level of recruitment in the symbols of the society as in Campbell (1983) regarding the effects of public opinion and sanction. Often these can be constructive, clearing land, building dams and the like, often destructive, as in political strife, pogroms, war, hysteria over economic ends as in the Tulip mania (Caldararo, 2013) or in fashion (Brafman and Brafman, 2008). How much of this is inborn and how much environmental?

Pinker argues against some of the extreme environmentalist positions that even Locke (often used as the starting point for the idea of the "Blank Slate" of human birth) realized there were "innate capacities" that "mediated human development." But Pinker specifically points out in his book, *The Blank Slate*, the following:

> So while Gould, Lewontin and Rose (Pinker's examples of such environmentalist views) deny that they believe in a blank slate, their concessions to evolution and genetics—that they let us eat, sleep, urinate, defecate, grow bigger than a squirrel, and bring about social change—reveal them to be empiricists more extreme than Locke himself, who at least recognized the need for an innate faculty of "understanding." (This quote and comments supplied by Pitchford, personal communication, 2003)

The nature of human understanding is certainly unclear. If all humans understood each other, there would be more efficient communication, learning, and, perhaps, order in the world. But the idea of "understanding" is often complicated by need, intention, competition, and features like physiological state and willingness or ability to respond. Often, people misunderstand each other due to a failure of attention or lack of skill in language or of the ability to reference common experiences underlying "words" and their meaning. Clifford (1988) refers to the work of Dilthey (1914) as a means of relating the "constructed realities" of individuals and cultures. Dilthey's "shared world" and idea of "common sphere" is made up of experiences people have had, which are structured in similar ways and made reference points (like the referents Deacon finds useful) that individuals utilize to reach common means of communicating. These concepts, "shared world" and "common sphere" are nothing more that complex conditioning schedules arranged in memory and accessed at appropriate times.

However, there must be the underlying physiology and developmental conditions on which such exchange can take place. As in the case of the sparrow, if males are isolated at hatching, they will still produce a rudimentary song, but if deafened at hatching, the song was even more incomplete than normal male songs. Thus, the song's instinctual basis is modified by self-reference in singing and the young male learns specific elaborate patterns from its parent (Bonner, 1980). Thus, nature–nurture is a complex system that is interrelated. Of significance is the usefulness of modifications of behavior. Depending on how changeable the environmental agents are, a modification may only have limited time value and new modifications may be necessary for survival. Here, forgetting and variations in retention in memory are important. This may account for some learning variations in vertebrates (Macphail, 1982) and here humans benefit from having no automatic generational transfer of learned behavior at birth, though some influences due to epigenetics and imprinting are assumed (Goos and Ragsdale, 2008). But, there appears to be evidence of variations in brain size that are due to seasonality, as in voles and shrews (Laskin, 1989), and considerable plasticity in cell size and connectivity in mammals and birds (Balthazart et al., 2010), while some have argued that seasonality can be a constraint to brain size evolution (van Schaik et al., 2012).

NO BLANK SLATE: LEARNING AND MUTATIONS

So, perhaps there is no "Blank Slate" but rather a complex of physiology; developmental process (including the movement of tissue [Trinkaus, 1969] and organs during ontogeny, digestion, heart beats, etc.); instinct, to some extent, to drive tissue interaction and cytochemistry in epigenetic tempo; the migration of neurons in the fetal brain and the process of connectivity selection that Deacon (1997) so rightly emphasized; as well as what we might call learning.

But Pinker makes a reasonable presentation, and a worthwhile investigation, for we should ask, "Why the belief in the Blank Slate?" No one can deny that the idea of the blank slate is popular, and has been throughout the past 200 and more years. It supports the idea of free will and thus bolsters some philosophical and religious argument, while it undermines other concepts, like predestination, fate, and aspects of reincarnation (Crapo, 2002; de Waal Malefijt, 1968). When we consider the argument Pinker is making against this idea, however—and now we go from Hull's complaint over Pinker starting with Locke—we find that our understanding of the nature of inheritance is abused. We recognize, for example, that combinations of genetic components and environment result in a variety of states of cognitive "abilities" and "disabilities," one demonstration of which is the many forms of idiot savant grading to genius, discussed in part I.

If we admit that there is little possibility at present for a "one gene," one behavior analogy, we resort to the idea of pleiotropy, though one might argue that some dominant and recessive genes can produce disabilities that limit perception or ability as well. In some cases, one gene can produce a physiological deficit as in albinism, which produces a perceptual "disability"—spoonerism. The individual, however, may respond with a variety of adaptations to this perceptual filter. On the other hand, the action of many genes with different degrees of penetrance and a variety of possible epigenetic developmental influences may be responsible for a variety of savants and idiots or genius and, of course, normality. Direct and indirect effects are described by Keller (2009).

Aspects of this are curious for understanding hominoid evolution. For example, the rate of accumulation of duplications has increased in African Great Apes relative to all other primates, producing significant disease-related copy number variation in humans (Conrad and Antonarakis, 2007); for similar reasons, Crabtree (2013) believes human intelligence is fragile and threatened. Barry (2013) came to a similar conclusion. Yet, while we find significant behavioral variability as well as tool using and claimed language acquisition in chimpanzees and gorillas (Patterson and Linden, 1981; Savage-Rumbaugh, 1986), we find little tool use in gorillas; yet, in contrast to expectation (given the concentration of duplications), we find more curiosity and manipulation of objects by orangutans (Napier and Tuttle, 1993).

Other cross-species investigations of cognitive phenotypes have studied deficits in attention, memory, and visuo-spatial discrimination using mice and humans. This included aspects of learning and the executive function. Most workers associate this function with the expansion of the frontal lobes in primates and decision-making. Other mammals have a prefrontal cortex that consists of two major regions, rather than three as in primates. The third region in primates, the lateral or granular, prefrontal cortex is believed to be unique in primates and concerned with "rational" aspects of decision-making. The other two regions are believed to be associated with "emotional" aspects of decision-making. Human behavior is so irrational, as we have been finding, that one wonders what is going on in this special "rational" region. Damage to this area in humans can result in reduced attention, disinterest in one's surroundings and lose track of organization, rules of order, and recognition of objects (Miller, 1999; Wallis et al., 2001). Also, stress can upset function, producing pathological behavior (Gamo and Amstan, 2011). So, one might say that, at any one moment, achievement and maintenance of rationality is fragile and fleeting. But as I have discussed in another part of this text, just as there appear to be many forms of logic and rationality, the nature of emotion seems also to be a construct, as expressed in humans, of culture (Reeve, 2015). Regarding decision-making, the role of emotion has become significant in recent years with the work of Damasio (1999), who has investigated the role of various brain areas in emotional responses and how these become integrated as homeostatic loops in what he termed the emotional or somatic marker hypothesis (Damasio, 1996). His approach is similar to that of Etzioni (1988) in that Etzioni demonstrates how emotions have been "played down" in behavioral research, especially in economics.

Truncated forms of the protein SRGAP2A, or the SRGAP2C in humans was also apparently expressed in Neanderthals and in the Dinisovans (Dennis et al., 2012). I have discussed the nature of speciation in *Homo* regarding these two hominid types of *Homo* in a recent publication and have reproduced here, as figure 4.2, a chart from that publication suggesting a revision of *Homo* classification to a form of anagenesis, arguing less speciation in hominid evolution (Caldararo, 2016b). The new find at Jebel Irhoud (Hublin et al., 2017) supports my view of a more inclusive transition from Archaic *Homo* to modern human. Some have argued that this mutation produced a new role in human brain function, yet we have little evidence, in behavioral terms, of this change. SRGAP2 leads to higher density of dendritic spines and longer shafts that are characteristic of human phenotypes than other mammals (Benavides-Piccione et al., 2002). Thus SRGAP2C is supposed to indicate neoteny in synaptic refinement. According to Striedter (2005), as brain size goes up in mammals, cell density declines and the extra space between cell bodies in larger brains is mainly filled by dendrites. As I have mentioned above, new discoveries in brain cells, especially the connectivity of astrocytes and glia cells, are changing our views.

It is difficult to make sense of other massive, highly conserved, yet rapidly evolving regions of the primate genome. It is believed that over 500 human-accelerated regions exist, along with a similar number of primate-accelerated regions, in comparison with 29 in mammals (Jones et al., 2012). Some estimates of noncoding human-accelerated regions are as high as 2,649 (Capra et al., 2013). Mutations in these regions may disrupt cognition and social behavior (Doan et al., 2016), supporting Crabtree's (2013) fragile cognition theory, with certain anti-progressive potential. Loss of genes promoting plasticity might not be damaging to the process of superorganism development but rather result in selection, which might mimic that behavior seen in domesticated animals, and technology might compensate for losses. In a way, this could be seen as paralleling data for ants with decreased brain size and increasing density (Riveros et al., 2012). Yet, over 500 regions are found in other mammals, including chimpanzees, but deleted in humans (McLean et al., 2011). Is it wrong to expect some significant behavioral change in humans, with all this genetic change, that would show up in the fossil record (e.g., rapid expansion into new niches, vast elaboration of tools, structures, etc.), or is this expectation simply a result of our materialist society? Or did these changes in brain function produce in early hominids, say Neanderthals and early anatomically modern humans the first intellectual faculties, producing philosophy and magic, religion and astronomy? Behaviors might have resulted that would not produce archaeological traces, like dancing, singing, rituals, and storytelling?

The reason why some anthropological geneticists have considered the analogy of the blank slate as an apt one is due to the incomplete penetrance of expected ability associated with IQ and other forms of cognitive talents. Since we theorize that there is a combination of genes, alleles, recoding events, and epigenetic developmental variations involved, it is not to be expected that abilities will be inherited in families, but rather in populations if they confer, as combinations, unique survival value over long periods of time. This is, of course, the position of Lumsden and Wilson (1981); we may assume that it is possible, and perhaps logical, that over the past two million years, selection for such abilities that we associate with hominid behavior can be expressed in features as obvious as different tool advances. But just as the "Creative Explosion" advocates propose that there was an explosion in cognitive ability reflected in art and technology some 50,000 to 100,000 years ago, we might similarly point to significant changes in behavior like the appearance of Oldowan tools at 2.6 million years ago, or the Acheulean at around 1.6 million, or the Levalloisian at about 400,000 BP, or blade tools. But this presses the idea of gene mutations, rather than discovery and learning, as motive forces, and we might then make the absurd argument that reading, or every invention from gun powder to the internal combustion engine, was the result of a mutation. But learning and use do have limited and specific roles in brain area variations in some cases, as with London cabbies (Maguire, 2000) that have larger hippocampus areas than average Englishmen. Lifetime changes in the brain have been well documented by Diamond (1988). Whether this is an artifact of sampling size or sampling procedure could be possible, but it parallels other examples (Striedter, 2005) and similar to reported changes in grey matter content of parts of the neocortex of macaques in response to changes in social group size (Sallet et al., 2011). It is similar to the Greenfield-Bruner thesis that maintained that schooling pushed cognitive growth to new levels, but Goody (1987) maintained that patterns of learning existed prior to the forms of complex society to produce similar results. He notes that in comparing literates and nonliterates, their references to behavior differ significantly due to the effect reading produces of a separate cognitive field or reference, just as in the use of hand-held computer devices today, both of which, reading and computer skills are learned but build upon previous platforms of tool use and changing the environment, or what Dawkins (1982) called the extended phenotype, as mentioned above.

Nevertheless, the interaction of learning and selection on genes must be intertwined, yet it is also obvious that general intelligence is a product of limits, both physical (e.g., expensive tissue) and social (e.g., war kill smart and less so, as does disease). As a measure of ability, this is a rather blunt instrument, originating in the work of psychologists like Charles Spearman, (1904) and purports to measure broad mental capacity that influences performance on cognitive ability. It is often referred to as "g factor." Macphail (1982) has discussed the problems in applying this concept across vertebrate species and the more intriguing difficulties when comparing animals with surgical removal or damage to parts of the brain. Problems in measurement led to criticism especially by L.L. Thurstone (1957), instead of a single "pool of ability" as Spearman had found, Thurstone discovered seven different primary abilities. Howard Gardner (1983) exploded this view of abilities into a scheme of "multiple intelligences." The lack of empirical evidence has undermined this theory and support for general intelligence has regained prominence (Waterhouse, 2006). An equally significant trait or skill is that of teaching. Were early hominids teaching tool making? We can see, from recent research, that teaching is not as rare as once thought. Crows engage in teaching (Cornel et al., 2012; Marzluff, 2013).

We still find a continuing trend where there is an understanding that bigger brains mean better decisions, as in the case of Alexander (1989), but decisions at what? To what end? If survival is the only determinant then most people have done rather poorly, dying "young" or perhaps that is not an elegant goal, but then many fail to produce offspring or those offspring fail to survive or reproduce. Even where the assumption is that bigger brains or vertebrates close to man on the evolutionary scale should "do better" at tasks we find exceptions, as Macphail (1982) shows, some lemurs outperform anthropoids at learning-set achievement, as do mink and ferrets and dolphins over marmosets and squirrel monkeys. Some reports in the existing data of comparative studies, as mentioned elsewhere, did not control for context and others did not control for physiology, as when Devine (1970) found that cebus monkeys (whose color vision differs from rhesus monkeys) scored poorly on tests where color was a component but scored well when all objects were a uniform gray (though he notes that interspecies differences were significant, and small sample sizes in tests produce uncertainty).

Perhaps the social IQ of humans is only effective within the superorganism of human society, but then, most people seem to perform poorly at that task as well. But what if our social intelligence was adapted to achieve performance in collective contexts and yet also be general enough to survive collapse of complexity as well? Here then, human intelligence and the brain's plasticity could be coupled with a cycling of complexity as a process of renewal, something similar to that seen in the social insects where, for example, a bee colony must regenerate itself periodically by swarming, new queens being produced and new locations created (Forel, 1930; Holldobler and Wilson, 2008).

The old cultural geographers like Ratzel saw a progressive change (1896–1898), as did other philosophers and historians before them (e.g., Vico and Ibn Khaldun), and thought there has been a more cyclical process of increasing and diminishing complexity. Gould (1988) addresses this historical recognition of time and change in the context of the development of the idea of deep geological time in the nineteenth century. This is because we find a forward and backward process of advancement. Human societies were seen after the Feudal Period in the west as no longer progressing directly, but rather advancing and contracting in cultural expression. We can hardly theorize that humans lose intelligence and then regain it, or as in the Maya (Post Classical) or English (post Roman), there is more to human cognitive expression than that encompassed by the individual. We lose information (as in the Mayan books) and yet can recreate it due, perhaps, to the general patterns of our experience or the psychic unity of mankind. We see this possible reinvention in Snorri Sturluson's (1179–1241 CE, Icelandic historian and politician) reinvention of euhemerism, which argues that the gods were once human leaders deified in time (a theory proposed by Euhemerus, a mythographer in Macedon in the fourth century BCE and cited in de Waal Malefijt, 1968), or the invention of the concept of zero, or of the wheel in both the New World and the old (Ekholm, 1946; Kaplan, 2000). This demonstrates the vast potential for human creativity and invention and is the best definition of the blank slate to my way of thinking. Civilization is not unilinear; neither is family ability.

If we reached a cognitive stasis some 100,000 years ago, what do we attribute it to? Cohn (1977) believed a general collapse in food availability could be discerned from the archaeological record after this time. He called this period the "food crisis in prehistory," yet Pfeiffer (1982) calls the period after 50,000 BP the "creative explosion." Gibbons (1993) and Self and Blake (1998) argued for a "bottle neck" collapse in human population after a volcanic eruption in Indonesia about 75,000 BP. Some, like McBrearty and Brooks (2000), as mentioned above, disagree and see a more gradual accumulation of products of creativity and need.

On the other hand, others, including Herskovits (1965) argue that there was definitely a reduction in ability that rises from the Aurignacian through the Magdalenian and then gives way to a crudity characterizing the Mesolithic.

Perhaps Cohn is right and the selective advantage that had created this flowering of human cognition gave way to a different kind of selection that did not favor a general advance in ability. But Herskovits argues that this crudity in art also appears as new materials become used and old materials are used in new ways.

My own ideas are expressed in a 1996 article titled "The HIV-AIDS epidemic: its evolutionary implications for human ecology with special reference to the immune system."[1] This work draws attention to the effects of disease and war, or general violence, against conspecifics.

We should consider the vast number of diseases, especially pandemics, that have assaulted humanity over the past 10,000 years. Some, including AIDS, threatened to destroy our immune system while currently the Zika virus not only is producing thousands of severely disabled infants, but recent experiments in mice show that it has the potential to sterilize males who are infected (Govero et al., 2016).

In this interpretation, domestication has had negative consequences as well as benefits and the short history of eusociality has been threatened time and again. Certainly, as we became more associated with living with the animals we domesticated, the results of the accumulation of septic conditions and pathogens jumping the species barrier to us, had numerous effects. Certainly, selection against those genetic phenotypes most susceptible to new diseases would take place as well as selection acting irrespective of ability as smart and able would die from disease. However, a stasis of brain size might have resulted, limited by its "expensive tissue," as Aiello (1997) argued that the selective advantage of brain size development appeared to "cannibalize" the size of the gastrointestinal tract in most primates, and that the sacrifice of the digestive tract was accomplished by a switch to more high-energy foods. Thus, a shortened gut led to big brains led to animals as foods to hunting, cooperation, social complexity, and modern society. As before, the main problem here is the fact of large brains two million years ago and they became larger but behavioral complexity did not. But, if a limit to brain size due to the expensive nature of brain tissue was reached, then it seems that neither brain size nor hunting or cooperation determine complexity in humans. If high-caloric food is a foundation for the evolution of big brains, why then do we not see trends in carnivores possessing big brains, with the possible exception of some of the canids (Finarelli et al., 2009, but see also van Schaik et al., 2012), when instead we find more rodents and primates?

Carnivores must engage in environmental mapping to hunt, many like canids do so in groups. If manipulation is a significant factor, especially in hunting, then we do see a trend in some cephalopods with the need to control their tentacles as primates control their hands. Cephalopods do have complex cognitive processing units, though these brains differ markedly from vertebrate brains (Shomrat et al., 2015; Young, 1971). They are capable of a variety of complex cognitive abilities as I discussed above briefly. Yet, there are more ways than one to get around the "expensive tissue" problem. In some birds, the brain size varies with neuronal number and density on a seasonal basis, not just in songbirds but also among food hoarders (Smulders et al., 2000). Thus, the selective advantage of the big brain in aiding in strategies for food acquisition and sequestering is balanced over time by reductions in brain volume costs.

Complexity is a late-appearing phenomenon and does not appear uniformly in human societies. It is certainly an anomaly in primates, and appears to be as unstable as it is in insects. Yet, this is a conundrum, if increased ability was under positive selection for more than two million years in hominids and produced the big brain, why was complexity so late, unless civilization is simply an epiphenomenon of social hunting at the end of the last glacial period? This returns us to the argument of Klein and Edgar (2002), and Cochran and Harpending (2009), that there has been a change in brain chemistry that is the cause of modern human behavior. Could a single mutation or a set of mutations be the foundation of this behavior, or could it be disease, as in the case of behavioral change in some animals infected with certain pathogens (a millennia-long St. Vitus Dance, Swedo et al. [1993] or mass psychogenic event Caldararo [2012b]), or, as Tim White (1992) suggests, cannibalism?

Many theories of the evolution of the brain have placed emphasis on primate models where males are the center of hunting or social group formation. These have produced some views where male power, competition, and sexual domination of females are the driver of this evolution (Alexander, 1989; Morris, 1969; Tiger, 1969). Dahlberg (1981) was among many who began to challenge this view and focus instead on the behavior of females in primate species and the variability of roles among surviving vertebrates (Hrdy, 1981). Dunbar et al. (1999) make a similar but mainly male-centered argument based upon females seeking to mate with the fittest males. Hrdy (1999) has argued that the mother–child relationship is at the foundation of human interaction, but unlike other theories of this relation, she argues it is not innate in humans and is the largely forms the foundations for human cooperative breeding linked also with allomothering. Here we could make a parallel to African cichlid fish whose uniparental care is a factor in brain size evolution (Gonzalez-Voyer, 2009). Hrdy builds on this (2009) to construct a theory of mind based on the infant's need to respond to stimulation from many providers, structuring a social format for understanding human action and cognition. This new focus on the woman's role in the evolution of the mind was long overdue, especially given the focus of most other writing has been on men, but when women did come into view it was usually as passive and responsive actors (Campbell, 2002) or the image was drawn from largely sexual perspectives (e.g., Joseph, 1996). For a rather mundane example, the ideas surrounding the string skirts of the Upper Paleolithic (e.g., from Gargarino) or the Neolithic (e.g., from Sipintsi) and Bronze Age (e.g., Grevensvaenge or Egtved) noted by Barber (1991) have usually been interpreted as textile means to entice men to sexual behavior as they are open at the back. A more practical explanation (Homerian citations for Hera aside) would be that they allow one to defecate or urinate without soiling the skirt. So perhaps there is more than one explanation, and not just a sexual one.

Sexuality might have played a role much different than people with Joseph's theories of women might imagine. Like bird species whose males lack a penis, the necessity of cooperation to achieve coitus is imperative and if the rewards of mutual sexual achievement provide sexual advantage, either for reproductive success or group solidarity (as seems probable in coupling in birds and their examples of social grouping) then the development and evolution of conceptual sexual rewards and identity in partnership would be a clear avenue. If this were true one would imagine that today humans would be doing a better job of satisfying each other. And yet there is evidence to consider that this was true before the Axial Age about 2,500 BCE if we can use Malinowski's data on the Trobriand Islanders (Malinowski, 1929) as a point of reference compared to the world denying religions that became dominant after the Axial Age. However, it is always dangerous to attempt simple, single factor explanations for any evolutionary or historical event (Boas, 1920; Simpson, 1949).

Hrdy (2009) and Dahlberg (1981) go far to extend the role of female humans in the evolution of the brain and culture. It is certain that most of the discussion in the past 200 years has focused on male problems, hunting, defense, maintenance of order, and so on. But it should be obvious that the condition in which females have found themselves, both as individuals and as a group has had a profound effect on not just the evolution of the brain but on language. Early female hominids must have recognized that their bodies grew to be different from males, and that they were liable for reproduction in pregnancy and vulnerable as objects for male sexual expression. How females mediated these demands on their bodies, at the control of males, must have placed significant adaptive value on communication and strategy. This must have been exceptionally so for those females who found themselves without interest in (or perhaps a horror at) giving birth and the process of body transformation it requires. The death of a female relative or member of the group during the birth process would have been a striking example of this danger. The same is true of sexual behavior for those premodern females who were not inclined, in both cases refusal to participate (have sex, become pregnant) would result in a deadly finality as we see from historical records (Coulton, 1925). If we include the physiological stresses of the female life as in pregnancy and child birth, for which there are no male equals, which Montagu (1953) included in his book

The Natural Superiority of Women, we might find logical foundations for the developmental brain differences.

Women who resist men have suffered the worst of all tortures and indignities from being buried alive to being cast out of society burned alive as witches (Coulton, 1925). Darwin (1871) may have been right that the model for animal domestication was the reduction of women to the slavery of men. One need not theorize a 2 my past of male domination to arrive at this scenario. Even if one posits a human society of a more Chimpanzee or Bonobo structure, or the egalitarian theories of early hunter gatherers, past matriarchies (Marija Gimbutas, 1989 or Bachofen, 1861) or gynocracy, the demands on women as opposed to men have likely been significantly different in the structure of perceptions and reality. Critiques of such ideas only amplify the concept of pressures historically delivered on women (Eller, 2000). While these ideas are interesting we can also see that there are differences in society in history where records show changes in women's status over time, as in Rome. We can also refer to ethnographic reports of the great Matrilineal Belt across Africa where women have generally higher status than in patrilineal societies (Stone, 2010). Yet an overwhelming amount of data on male/female relations in humans seems to justify the idea that Darwin presented.

I am not speaking here of the structure of male and female brains, though these become different in the process of ontogeny (Brizantine, 2006; Darlington, 2009), the core of the difference that could give rise to a complex human consciousness may be found in the nature of being woman, of parrying off or mediating the biological demands of a more powerful male. Obviously we find both equality and asymmetry in Primates, the one often cited are the gibbons and siamangs, the other in Gorilla and many monkeys, but in the *Lemur catta*, as Hrdy (1981) notes, we find females dominant. Primate social relations range widely. Central to many current theories of the evolution of the mind and its complexity in vertebrates is defined by cunning and deception, what greater need is there to possess such talents than to avoid being the pleasure of a group of male hominids or to be translated into a baby-making machine. Some feminists have described marriage and conjugal love as serial rape where the female must constantly negotiate with the male his right of access to her body (de Beauvoir, 1952). Deacon (1997) and a number of others (Tucker, 2014) have given marriage (pair bonding, certainty of paternity) a central role in human evolution, but the variety of human marriage (polygyny, polyandry, divorce [Stone, 2010]) indicates that marriage is simply another form of domestication, but of women (see also Firestone, 1970; Greer, 1970). The ways that men, especially, have availed themselves of sex via serial marriage, divorce, and prostitution are little different than sexual behavior in chimpanzees, and paternity is uncertain unless women are made prisoners (some cultures have come close to this).

If Tim White (1992) is correct and cannibalism has been an important human activity since the Pleistocene then this is even more important to the understanding the slide in ability. But my own idea here is that as populations became denser and disease played a more important role in selection for immune genetic response and the random effects of slaughter due to war, that a generalized level of human selection for cognitive abilities was established. This combines my views with White's and Cohn's. I also think, however, that the reorganization of the human brain that took place (discussed above in part I and Striedter, 2005) in the Pleistocene gave rise to a limited maximum development and that a relatively fixed genetic plasticity of cognitive ability was achieved which arrived at a general level as the most adaptive. Not brain size, but brain cell type and connectivity efficiency as noted above regarding glia cells and astrocytes. This is what we might say was the cause of the stasis in human brain size growth and slight decrease.

While a shuffling of genetic influences occurs throughout the human population, the outcome in individuals always varies between idiot savant, generality and genius in a, say circular relationship maintaining a general ability as that most advantageous for human society. The individual may suffer in such a balanced system of polymorphisms, and evolution of ability may be stymied or have reached a maximum given the expense of complexity, but the result cannot be doubted would be efficient for society to maintain, given modern stresses mentioned. The complexity, note references here to the vast increase in repeated sequences and copy number, for example, would lead to reduced efficiency or nonfunctionality and what is described in law as madness or insanity. Selection by religion or by conformity to urban life also have most certainly had a generalizing effect, Inquisitor Tomas Torquemada was eloquent in his perception of the too curious or clever individual as devil-like (Longhurst, 1962). We see a similar aspect of jealously and spite which can destroy a bright mind in societies from Dobu (Fortune, 1932) to the Mediterranean (Theory of Limited Good, Foster, 1965). In fact, Joseph Campbell (1991) mused that human sacrifice in traditional cultures was often not only the most "aesthetic" looking individual but the most brilliant. Thus the idea of society as a selective pressure to maintain a balanced cognitive ability in humanity has a variety of factors involved (Gerringer, 2006).

The debate over nature and nurture continues, both regarding Pinker's work and that of E.O. Wilson and the idea of universals in human behavior and culture. A part of this appears in Donald Brown's book *Human Universals*. Here we find a number of reexaminations of anthropological studies of the late nineteenth and early twentieth century attacked by later researchers and found not only deficient, but part of a grand scheme by the "dogma of relativism." The intention of the author is to argue a kind of conspiracy by anthropologists to impart to man a certain plasticity. This idea seems to undermine concepts of how we understand humans to differ from animals and yet is based on the idea of sociobiology. If we refer to these, starting with attacks on Mead by Freeman, we find that each of them is by a person who arrives later, often decades later, to find contradictions in the "special" nature of cognition described by earlier pioneers. The problem is generally ascribed to either a bungling misunderstanding by the original researcher or downright fabrication (Caldararo, 2004c). The "quest" Brown (1991) is on here seems to be to identify what the original human was like, a proposition which appears so subjective as to be impossible. He is right that anthropologists have been attempting to understand what it means to be human by examining the cultural variations, but this is no conspiracy. But by studying our diversity of human being, we might be able to imagine a past with the help of the archaeological record.

There is no sense in his criticism that time and contact has produced the parallels with global Westernizing culture, rather the attacks, against Whorf, or Benedict, or Mead, are always described as if the new researcher should find exactly the same features analyzed in the texts and notes of the original researcher. That person ascribes any variation to failure. The general conception of these criticisms is that primitive societies should be fossils, unchanging and unaffected by modern society. I have detailed the way Freeman constructed his "findings" against Mead in my 2004 article, but a short discussion of the Hopi idea of time is another interesting example (Caldararo, 2011). In an attempt to sort out ideas of origins and modern, Montagu (1968) wrote a whole book on the idea of the primitive, which Hsu (1964) found to be quixotic at best since he argues that looking back from complexity one tends to find difference as "primitive," or less than. Redfield (1941) substituted at times the idea of rural–urban contrasts; Herskovits (1965) sometimes used nonliterate, and Goldenweiser (1922) ideas of small, simple or isolated. To Hsu these all possessed the connotation of inferior, backward or savage, implying "nonhuman." A problem here is not so much the hubris of civilization, but the demands of certainty. Montagu addressed this pressure with Darling (1970) in their book *The Ignorance of Certainty*, respect and expertise demand knowledge be expressed definitively. The public wants to know what the first human was like, what did she/he look like, do, say and what did their genes look like?

The difference between science and belief is that we argue and compare. Darwin produced his Theory of Natural Selection and then spent nearly two decades trying to make sure he was right (Darwin, 1887). Wilson produced Sociobiology in less than five years and called it a law of biology. To read the book, which it seems few people do, most become then either "sociobiologists" or "anti-sociobiologists" as if we were taking on the mantle of some minor religious sect. Many then seem to not bother to read Darwin after the experience, and one finds a torrent of absolute statements about behavior that cannot be substantiated or disproved, it is almost as if Wilson produced a work of faith for some people. Caplan's (1978) compilation of essays is a good example of the fall out of this debate. After reading *Sociobiology: The New Synthesis* (1975), one realizes what a masterpiece it is, certainly on the level of Darwin's *Origin*.

The main problem is the absolutism that is associated with Wilson, though his work is not only reasonable but also supported by a mass of scientific evidence. Wilson's position on IQ is clear:

> The hereditary factors of human success are strongly polygenic and form a long list, only a few of which have been measured. IQ constitutes only one subset of the components of intelligence. Less tangible, but equally important qualities are creativity, entrepreneurship, drive, and mental stamina. Let us assume that the genes contributing to these qualities are scattered over many chromosomes. Assume further that some of the traits are uncorrelated or even negatively correlated. Under these circumstances only the most intense forms of disruptive selection could result in the formation of stable ensembles of genes. A much more likely one is the one that apparently prevails: the maintenance of large amounts of genetic diversity within societies and the loose correlation of some of the genetically determined traits with success. This scrambling process is accelerated by the continuous shift in the fortunes of individual families from one generation to the next. (Wilson, 1975, p. 555)

Increasing numbers of birth defects, errors in transcription and translation, epigenetic variations, duplications and increased copy numbers also complicate this "scrambling process." The raging complexity leads to a question: how can the expansion continue without a dramatic increase in entropy and system collapse? While some place faith in IT and robotics, we might find a future looking more like the movie *Forbidden Planet* where the intelligent life forms unleash the most irrational needs by the creation of an automated planet. This reminds me of the confusion that people with Alzheimer's and some other forms of dementia whom I have spoken to, find it difficult to discover what is reality or dream. Since the brain creates reality from external stimuli, one wonders how the experience of Virtual Reality will affect the process of reality formation as it has evolved, and what results that interaction of realities will have on the phenomenon of mind and its ageing. Put this into a consideration of the perversion of technology in the movie *Zardoz* where the inequality in wealth and information creates two castes, one a pleasure seeking elite without real function, dominating a mass of warring people left in ignorance only to produce diversions for the elite. Reality again is distorted, as is humanity's relation to the natural world (one could substitute any number of similar scenarios: the film *The Matrix*, for example). Similar "solutions" to problems like longevity are primitive in conception: to increase human life span will only place more stress on an already burdened planet, and ideas to populate the stars or the moon[2] are more stage-managed hubris than likely futures. This brings to mind the conversation of Coyote/Saynday (Marriott, 1947) over the problem of too many people in the world; the wisdom of this myth seems lost to us, as if the planet has no end of room (Lynch and Roberts, 2010).

Lovelock's ideas of Gaia or a dynamic system of the planet and its resources seems too quaint to matter in our time (Lovelock, 1988). Given the attitude of most people in the world regarding the use of the planet and its meaning, things are rather dire. For example, Niebuhr summed up the philosophical picture of the Judeo-Christian community in the despair following the Second World War thus:

> The only hope of men lies in the miracle of divine mercy and the only promise of the kingdom of god that men can cherish is the promise of a radical, cosmic revolution which will substitute for the present world with its natural laws as well as its social evils a new heaven and a new earth entirely different from the cosmos as it now exists. Every human effort is not only futile but damnable. (Niebuhr, 1957)

Should other animal life on the planet have the consciousness to recognize the threat that humanity under this ideology of world-rejection holds, they might object, or fear complete destruction. But every major religion today, with the exception of Confucianism (if it can be so classed according to Weber [1959] as a religion) holds a similar world-rejecting attitude as the Judeo-Christian tradition (Bellah, 1978). Therefore, life on earth is endangered, indeed. To the quote used by Pinker (2011) as a title to his book (from Lincoln's First Inaugural Address, March 4, 1861) Niebuhr added, "Is there no continuity between the divine mercy and those angels of man's better nature which struggle with the demons of the jungle in his individual and corporate life?" That ability to motivate groups and to inspire them to terrible results, of which religion is a manifest expression, may be the quality of mind that ends the human career.

Humanity has not evolved to the point where its creativity can be controlled by a morality and since no one can agree on what is moral, the price of survival looks quite expensive indeed. Usually concepts of morality reflect values of particular culture histories, as for example, Wright (1995) in which philosophical idealism fails to match the reality of actions.

It may be that some individuals act "morally" according to specific codes, but when the median social value arrives at a Thirty Years' War or colonial or Nazi ideology, the mass behavior reflects that concept of morality with all its terrible consequences of self-righteousness. Peoples' values are the result of enculturation processes, and these are complex conditioning schedules that are quite malleable. What we eat, whom we eat and how we treat what we eat are all defined by these situational dialogues that happen generationally and across generations. What we consume and how we consume it is embedded in our cosmological framework, and as in the case of the AIDS epidemic (Caldararo, 1996), the consumption of bush meat apparently unleashed the Human Immuno-Deficiency Virus on the world (Pepin, 2011), as agriculture spread the plasmodium that causes malaria (Carter and Mendis, 2002).

Yet ideas of benefits that can be derived from a study of past diets or behavior may be misplaced. Often the archaeology of the past is interpreted in the context of the present as when people speak of the "Paleolithic diet." And books have been written on this topic to claim that they have discovered diets that can achieve longevity and health (Cordain, 2002), but there are many factors involved in longevity and health, including genetic factors (Ungar, 2007) and cancer scientist Mel Greaves (2000), supported by cross-cultural research by Fabrega (1997), has argued that hunter-gatherer humans of prehistory were adapted to scarce resources and this resulted in short life spans and susceptibility to chronic ailments, resulting in younger and "fitter" populations. Strength and endurance would be at a premium, and Bogin (2001) describes quite varied diets among contemporary hunter-gatherers. Some ideas about diet in hominid populations have been questioned by reexamination using new methods, as in the case of the Neanderthals who are often characterized as being mainly meat eaters, yet now data supports a wider diet of plant foods (Henry et al., 2014). We have to be concerned with what kind of health we are considering (Katz and Meller, 2014). A very active but short life? A long life characterized by low-impact scavenging and gathering or horticulture? A long life with active late years of senescence? We also should guard against idealizing periods of the past (Zuk, 2013). Was morbidity of the Paleolithic comparable to the lives of millions who toiled on manors as serfs, or on plantations in the colonies, or in the noisy and dangerous factories of the Industrial Revolution?

Gurven and Kaplan (2007) survey contemporary populations on longevity factors and dismiss archaeological research on prehistoric hunter-gatherer populations as being too much in variance to contemporary living populations. They recognize that historic and contemporary hunter-gatherers were both stressed by contact with agriculturalists and complex societies and at the same time such contact modified their exploitation of the environment and changed their living conditions. Yet, archaeological data (e.g., Caspari and Lee, 2004; Cutler, 1975; Weiss, 1981) found prehistoric populations of hunter-gatherers living 15 to 20 years less than contemporary or historical populations. They conclude that something is wrong with this data, that it cannot be an accurate sample of the past as it would cut short the ability of groups to sustain and transmit knowledge across generations. This might be true, in fact, such a reduction in gathered life experience could have had a significant effect on information transfer and retention and it could well explain the fact that sedentism and complex society appear so late in human history as the life span appears to expand.

A number of people who have studied fossil human remains, as early as Weidenreich (1939), have found that the evidence indicates that few if any died of natural causes. As Wells (1964) noted, "The vast majority of skeletons, even when showing gross pathological changes—which most do not—seldom reveal what the person died from." This is a conclusion seconded by Ortner and Putschar (1985). The sample is skewed, as Steinbock (1976) remarks, by the fact that three-quarters of all archaeological skeletal material consists of skulls only. Violence, accident and perhaps infanticide or senilicide as reported in Eskimo (Inuit) by Van Hoof (1990) seem to make up most all cases. This would make determining a "natural life span" quite difficult. Nevertheless, if one were to focus on diet alone the Paleolithic diet would appear to doom one to a short, but energetic life.

The relationship between aging and dementia is clearly established as is certain diets with diabetes and dementia (Strachan et al., 2010). Prion disease is associated with certain foods and means of preparation (Prusiner, 2001) and cancers can be found more frequently in certain populations with certain diets and behaviors (Greaves, 2000). It may be that genes for longevity and those for different survival strategies are at odds, though there is evidence that at least one longevity gene may be associated with enhanced cognitive performance (Dubal et al., 2014).

The problem lies not with humanity, but with how ideas of human nature are reinterpreted. It seems to me obvious, and that is why discrimination and racism are closely related to those who promote ideas like the ones in the book *The Bell Curve*. The concept of inheritability of genetic ability was the core of Eugenics and Francis Galton's theories (Montagu, 1964), and when people read of a link between genes and human behavior they seem to come to Kantian conclusions on the nature of race. Humanity has experienced a history of competition shaped by war for almost the entire time we have lived in sedentary communities. The idea of difference is part of the heritage of that process. Behavioral traits of avoidance, repression and domination may simply be deadly to our survival.

The problem of race, as Ashley Montagu noted, is man's most dangerous myth. And we are caught between the ignorance of certainty and the fear of the other, so let us look at race as one of our present great dangers in Montagu's opinion.

NOTES

1. See also Niccolo Caldararo, *The Anthropology of Complex Economic Systems: Inequality, Stability and Cycles of Crisis* (Lanham, MD: Lexington Press, 2013).

2. Newt Gingrich's, for example. See "To the Moon, Newt!" in *Slate Magazine*, http://www.slate.com/articles/technology/future_tense/2012/01/newt_gingrich_s_moon_colony_and_mars_plan_.html, accessed March 20, 2017.

Chapter Ten

Racism as a Human Disease

Perspectives change our understanding and our position in the world, and what we are considered to be by that world defines our understanding in many ways. Franz Fanon put this idea in a redefinition of history: "That same Europe where they were never done talking of Man, and where they never stopped proclaiming that they were only anxious for the welfare of Man: today we know with what sufferings humanity has paid for every one of their triumphs of the mind" (Fanon, 1961).

It sometimes occurs that the accidents of history rob the world of great architecture, music, art and literature. Antenor Firmin's 1885 text on the equality of the human races, a monumental contribution to our understanding of race and racist philosophy written at the height of the colonial atrocities and the construction of the Supremist Theory, disappeared without note. Its loss was a tragedy to scholarship and the conflict of ideas and nations.

Firmin, born on Haiti in the wake of the revolution that created the first black post-colonial state, was educated as a lawyer in the schools of Haiti and traveled to Europe as its ambassador just before the American Civil War. He recognized the power of white supremacy was aimed at crushing his nation, and Southern senators and plantation owners called for invasions and blockades of the island as well as seizure of the property of its population to indemnify the lost value of its once-enslaved people.

EUROPEAN EXPERIENCE

Imagine the feeling Firmin must have had to be faced with a world so arranged against him. While other representatives of colonized peoples had come to Paris, Firmin was alone as a living threat to the status quo, not only due to Haiti's successful revolt, but also because of the reaction against the French Revolution, its ideals and the uprisings in 1848 and 1870. The French Revolution had called for freedom, liberty, equality, and fraternity among the people of the world, only to later recant and limit it to Europeans. He had to suffer the insults to his person by people his intellectual equals (and often not so equal in ability). Franz Fanon (1967) experienced this same contrast of ideal and real and, as a physician, came to describe racism as a disease created by an ill society. His vision of the demise of this illness was as catastrophic as in an ill patient, and the world wars a consequence of the illness. Kenyatta's (1938) experience of the contradictions drove him to study developed western societies (including the Soviet Union) and led him to conduct an ethnographic study of Western society.

It is amazing to realize that if one looks at a map of the world today, or any part of it outside of Europe, you notice that almost all the national boundaries are derived from colonial treaties on how to divide up the world. They seldom represent whole peoples or cultures, rather the opposite where peoples, for example, the Maasai are divided between Kenya and Tanzania. Even given a map made in 1970, there have been few changes, the division of Eritrea from Ethiopia, Southern Sudan, but largely the divisions of the world still are those imposed by European colonialism and non-Europeans still have to live and deal with the problems colonialism has created (Caldararo, 2015b).

Firmin was quickly recognized as a uniquely talented lawyer, diplomat and scientist. His interests, like those of his contemporaries, were widely spread over the sciences, arts and the history of humanity. He was elected to the Societe d' Anthropologie de Paris where he met and discussed the scientific issues of the day regarding humanity, with the leading scholars of France. These men were mostly convinced of the theory of Polygenesis and the inequality of man as defined in the racist text of de Gobineau (1853–1855). Firmin, however, preserved and amassed a mountain of evidence to combat this theory. When he did rise to debate the racist pronouncements of men like Broca he was attacked as a mulatto whose European heritage gave him the ability to understand and contribute. Shocked by this charge, unsupported by any proof, Firmin decided to quietly record the addresses in the Societe and wait to publish an analysis.

Drawing on Darwin's work and thousands of other scientists and the record of history he constructed one of the most comprehensive refutations of a theory in the past 100 years. Criticizing philosophical nonsense like Hegel's argument (1830–1831) that nothing of importance happened in Africa and that the "Negro" represented the state of the wild man, untamed, or Immanuel Kant's ideas of inequality patched up by Plato's political tracts, Firmin relies on the solid evidence of science and examined historical source. And even Kant's central thesis, seems borrowed from Parmenides (a pre-Socratic philosopher of the sixth century BCE), who was "the first thinker to set up a fundamental opposition between the senses and abstract argument or reason" (Lloyd, 1979). As Vico and Ibn Khaldun before him, Firmin dismantles the bias in writing in history to uncover foundations of a science of anthropology that exists today and his work stands like Boas' in a realm of clear thinking where all assumptions are examined and made to stand the light of day.

DEBATE OVER MONOGENIST AND POLYGENIST ORIGINS OF MAN AND EGYPTIAN CIVILIZATION

Anthropology, as a science, could have benefited from the translation and publication of this book 100 years ago. With Comte, Montesquieu (whom Evans-Pritchard favored as a founder), Theodor Waitz (to a lesser extent Tylor, but certainly Adolf Bastian, in his idea of the Psychic Unity of Mankind), and Pritchard, Firmin could be considered a founder of contemporary Anthropology, except his influence was limited. Firmin is modern in the sense that he rejected higher and lower races and questions the idea of progress (he would better be described as an empiricist rather than a positivist as he titles his book), a concept Tylor accepted, as did most of his contemporaries.

In Fluehr-Lobban's (2000) review of Firmin's book, her criticism of Tylor is typical of contemporary hindsight regarding early work in anthropology. Firmin noted that Tylor, like many others, was working with the reports that were available and attempting to develop a scientific means of treating sources to check for fact versus fiction. If racists are the only ones publishing cranial studies, then that is your database. In fact, Firmin shows how the scientists of his time could not agree on what was a race, nor how to define race with respect to the concept of species. Montagu (1964) addresses this issue as well as does Gould (1996b). This confusion is traced to Kant by Stocking (1988) and we can see it present in Leakey's (1934) attempt to produce an evolutionary tree including all primates, living and fossil. The thick stem of the "Neoanthropine Hominidae" at the leads to branching of contemporary populations listed as races: European, Hamitic, Negro, Mongoloid, and so on. Such models can be useful in contrasting the long evolutionary past of trends in biology, but when projected into the present they give the immediate impression of distinct populations whose traits are established over time and can be differentiated from each other clearly as if we can ignore clines of traits and the history of human migration and hybridization. Human populations show gradients of genetic information, not discrete geographic demarcations or groups (Serra and Paabo, 2004). The complications of understanding social groupings and barriers to gene flow produced by social sanctions and discrimination are difficult to quantify, as in work published in India (Bamshad et al., 2001; Serra and Paabo, 2004), where gene flow and the distribution of traits confounds caste distinctions.

In the debates of the racist polygenists and monogenists Firmin took quiet note of the passion of the argument and his text reflects both a devastating critique of de Gobineau as well as a thorough refutation of the points of other scholars like Bory de Saint-Vincent, Broca, Morton, Agassiz and Quatrefages. These theories were confused to Firmin by association with religion and ideologies, as in the case of some monogenists who argued that different races were immutable, fixed and defined by their creation and history, while some polygenists believed races could and had mixed, producing viable progeny. What is of interest, then, is why most all the works of the proponents of polygenesis or monogensis, racist or otherwise have appeared in English translations with the exception of Firmin and Bastian. Fluehr-Lobban (1998) has suggested that one reason for Firmin's obscurity and the failure of his book to be translated and printed before is the fact of his political activity. But this does not apply to Bastian, the other champion of equality.

But Firmin, like Ashley Monatgu after him, takes on the physical anthropological problems as well as the medical ones, especially the canard of miscegenation resulting in monsters or sterility. He also attacks concepts of intelligence and of degeneration. His knowledge of history outshines the ignorance of his opponents, yet Firmin's concerns were universal regarding ability and not affected by the nationalism of the time. He was struck by the European tribalism and caste system that exploded in revolutions (e.g., 1789, 1848, 1917) and European tribal wars (e.g., War of the Spanish Succession, Seven Years' War, etc.). He does review the history of African civilization and archaeology, coming to the conclusion nearly 100 years later, as do Diop (1974), J. Desmond Clark (1970), and Keita (1992), that ancient Egyptian civilization was a product of Africans. Herskovits was perhaps aware of the Firmin's work when he engaged in his ethnographic work in Haiti and collaborated with Jean Price-Mars, but there is no evidence he ever saw or read Firmin's book (Fluehr-Lobban, 2000). Strangely, Barzun (1937) praises Firmin as among the "imposing group of intellectuals" "distinguished in some profession" who have undertaken to write on race but appears to criticize Firmin for not understanding French ideas and producing a work that mainly "vindicates the Negro race to which the author belongs." One is left with the impression that Barzun read Firmin but felt his book was mainly political. Since my copy of Barzun comes from Cora du Bois' library and she supported the Herkovits's work in Haiti, one might expect she had some contact with Firmin's writing, yet there is no evidence of it.

Unfortunately today some geneticists are making similar arguments where a few bases in microsatellites are proposed to distinguish groups of people as races. Like Broca, these theorists throw out the entire history of species determination and rest their case on a new methodology. I have opposed this approach in several cases, one with Neanderthals and more recently with Denisovans (Caldararo, 2000, 2003a, 2016b), for not only do distributions of haplotypes cline, as do traits, but they produce complex geographic distributions (e.g., see Basu et al., 2003). Broca's bias charged the theory of polygenism and caused him to overlook principles of biology that he applied to all other species, specifically within group variation. Today those who sell tests to determine people's "heritage" or "origins" fail to note that their methods obfuscate the fact that haplotypes and microsatellites cline just as phenotypic traits do, someone may have one sequence which is shared by 80 percent of people in a certain area today and another which separates them from those individuals. But the main problem is the idea that nationalities or races have been stable for thousands of years or that there are "homelands" they can be associated with in specific areas. Is a French person today, who is part of a nation state, the same genetic relative as a person living in Lange D'oc in 1000 BP to someone living in Brittany among Celts in Roman times or before with the arrival of the Beaker People? Neanderthals? Anatomically Modern People? The same goes for identifying people by haplotypes as Nigerian or Fulani today. It ignores history and movements of people but also produces ideas in the public mind that peoples, nations, races have always existed as they are socially constructed today.

Opposed to Polygenist theory was the Monogenist theory derived in part from the Bible. But characteristic of Firmin's approach, he did not argue as a partisan of monogenist views either, but from a scientific basis of balancing the data and seeing where the preponderance of evidence lay. He praises the work of Friedrich Tiedemann (1836), who also worked in an empiricist vein to demonstrate that those who argued that Africans had smaller brains than other people were wrong. We still find people today who believe that racism is inherent even though research, especially with children and the developing brain have demonstrated this is false, most recently in work by Teizer (2012). In fact, what is often referred to as inherent racist tendencies, especially in groups, can be associated with what Lev Gumilev (1990) described as intensive cultural identities, enculturated under specifically harsh conditions.

THE NEW WORLD, HAITI, MISCEGENATION, AND CULTURAL RELATIVITY

Firmin analyzes the contributions of various peoples in Europe and elsewhere to forge an understanding of the relative values of cultural products used by Supremist theorists to demean non-European cultures. He also utilizes the examples of the achievements of the people of Haiti, especially after the revolution, a subject most Europeans of the time were completely ignorant of and to this day is poorly known or acknowledged. Firmin describes an anthropology that is as broad and curious as it has become today, and presages the advent of the Four-fields wholistic approach.

But Firmin finds that the debate over race is mired in a religious struggle and shows how certain scientists, like Broca and Georges Pouchet, took to the Polygenist theory of human origins initially due to the religious context of certain supporters and arguments of the Monogenist theory. The conflict of religious belief and scientific view was confused in this debate. Broca's views changed from creationist polygenism to a polygenist evolutionary position by 1878 (Langham and Oldroyd, 1983). Louis Agassiz maintained a strict creationist view, but even after the publication of Darwin's Origin and Descent, a scientist of the stature of Henry Fairfield Osborn could publish a major work of polygenist evolution as late as 1916.

Firmin immediately saw Broca's approach to human beings as a departure from the methods and theory built over a century for the species concept for all life. Broca had argued that with man the scientist could forgo the usual attention to details and look to the more obvious, phenotypic traits. This attitude led him to argue that marriages of members of different "races" would lead to infertility, monsters or worse. Firmin attacks this loose approach with organized vigor, showing Broca's argument to be biased and illogical when confronted with the evidence which Firmin could provide from his own Haiti.

Not only does Firmin face the idea of inequality of origins a basis for judging the products of different cultures, but also he finds that foreign cultures are often regarded as high or low based on the degree to which they depart from possessing or converting to Christianity. Pouchet and Renan deny the ability for equality on stilted interpretations of Christian dogma where morality is defined as a component of Christian religious life in stark contrast to the violence and depravity of the Wars of the Reformation, the massacres of the peoples of the New World and the inhumanity of treatment of non-Europeans by colonial governments and businesses.

Incredibly, Firmin constructs a theory of relative cultural norms what was quite parallel to that produced later by the American school of Anthropology. While using the hypocrisy of European society as did Kenyatta some 40 years later, as a vehicle to compare behavioral norms in European and non-European societies, his approach is clearly one of cultural relativity similar to that defined by Ruth Benedict and Margaret Mead in the 1930s. Firmin rejects the fascination of anatomists with cranial studies as a means to divide human groups into different species, predating Weidenreich's (1949) criticism of the practice by some 60 years.

The only major issue Firmin does not address is the inequality of women. That issue came to the fore during the 1870 Paris Commune and then awaited a later date for progress, both in work by feminists in Europe and the United States, though Simone de Beauvoir (1952) notes that the treatment of women as property and things without rights continued into the twentieth century with arranged marriages still common and even clubs for conducting such exchanges. Mead and Benedict as well as many feminist anthropologists in the 1960s and 1970s pointed out inequalities across cultures (Reiter, 1975) as well as culturally defined aspects of gender performance. Hrdy (1974, 1981) addressed a varied of concepts of gender in Biology that were unwarranted and verged on male chauvinism. Dawkins' (1989) and Trivers' (1972) discussions on parental investment are so influenced by such thinking that they largely produce stereotyped versions of primate behavior as well as birds, yet fail to notice that most birds lack a phallus and the interaction of male and female must be cooperative for the sex act to succeed, perhaps a basis for the more attentive male investment in young (Brennan et al., 2008; Wesotowski, 1999). Yet today we still have not seen great changes in hiring, pay and health access on an equal par with men for women. While major studies of the female brain have been made as mentioned above, we find ideas of inequality presented as biology in some works (Annis and Gray, 2013). Though this may be due to a misunderstanding of how ontogeny and hormones affect conductivity and the development of the organism to integrate stimuli into communication that varies in certain developmental settings (Darlington, 2009), it does not imply strictly male and female means of communication as Mead (1949) demonstrated so long ago. Luckily, other voices have addressed the structural social barriers than require modifications to achieve equal opportunity (Bohnet, 2016).

Since Firmin's day racist ideas have multiplied with as little scientific support as in his day as well as a greater understanding of human diversity and how the idea of race is a cultural construction and not supported by science. Some people argue that they can sense difference in others, and there is some evidence for this as in Wedekind's (1995) t-shirt experiment with the human leukocyte antigen immune genes. Female humans were apparently able to smell genetic differences in the males who wore the shirts, but the implication was that they were attracted to difference not repelled by it. This parallels findings with birds where discrimination of differences in male appearance indicates variations in fitness perceived by females and is a form of sexual selection (Velando et al., 2006). The ability to discern disease in conspecifics and others (as in dogs being able to sense cancer in humans, see Ehrmann et al., 2011) has some support and especially in insect societies (Caldararo, 2012b, 2015a). Santos et al. (2010) argued that they had found that children with Williams syndrome lacked the ability to form racist beliefs and cited studies showing that early in development children discriminate against other ethnic groups. The implication in both cases was that there was a gene for racism and anti-racism. But what the cited studies found, instead, was that children fear the strange and unfamiliar, noted in studies by Kagan (1970). But there is no evidence of a genetic basis of race, and fear of the Other is learned—whether it be fear of race, religion, or political affiliation.

FIRMIN IGNORED OR FORGOTTEN: WHY?

It is curious, however, that since the translation and publication of Firmin's book in 2002 in English, and the review of his life and work by Fluehr-Lobban in 2000, there has been little interest in either the book or the author in anthropology. Fluehr-Lobban (2000) argues that the reason the book was ignored was its "revolutionary premise." But there seems to be more involved. A similar voice was that of Jomo Kenyatta, whose book *Facing Mt. Kenya* in 1938 was met with a wave of criticism by social scientists and since World War II, post-Modernists, for his criticisms of the Soviet Union and his promotion of nationalism and nonaligned independence (Berman, 1996). Firmin was also a nationalist, but also with an international view as Kenyatta, but across Central and South America and in Pan African endeavors. Haitian freedom fighters fought with Simon Bolivar to help liberate Columbia, Venezuela and Peru, and of this Firmin was proud and vocal in his writings on colonialism.

One other factor is of importance and that is the effect of pro-slavery plantation investors who profited from inequality and racism. Firmin describes the process of organized degradation that was colonialism. The massacres in the Congo by the Belgians, those in Ache by the Dutch as horrible as they were do not out do those of the Spanish, French and English. Chiu Hsin-hui (2008) analyses the comprehensive process of colonial "domestication" of a people in Dutch Formosa that chillingly rings across the globe in other studies of colonial terror. Yet today's headlines of the struggle of native peoples and local populations against global interests seem reminiscent but lost on the world media (Caldararo, 2015b). Some historians today even have praised the colonial process as bringing civilization and technology to non-Europeans (Ferguson, 2002, 2011).

A book like Firmin's would not be received well by the individuals who profited from slavery and colonial investments in his time. Other efforts created an alternative anthropology in the nineteenth and early twentieth centuries that I have documented elsewhere.[1] But the suppression of their voices due to their revolutionary behavior, their criticism of colonialism and racism delayed a more effective anthropology from emerging.

A less conspiratorial explanation might be found in the very language that the book was written in and never translated into another. De Gobineau's was translated, and as a result was the subject of numerous reviews, including Boas' (1918) condemnation. It was given dramatic focus by the German composer Richard Wagner in his magazine, *Bayreuther Blätter*. It was translated in 1856 and again in 1915 into English and therefore made available to a wide audience. Firmin's book had no such support; it was ignored in France for the very reason of its contrast with de Gobineau's message. As de Gobineau stated in a letter (1933), his book was based on "a hatred for democracy and its weapon, the Revolution, which I satisfied by showing, in a variety of ways, where revolution and democracy come from and where they are going." He was an aristocrat who pushed his religious beliefs, racist and class associations. Opposed to him was Firmin, a Haitian revolutionary, a man of reason, committed to the law and democratic beliefs. There could be no greater contrast.

It is unfortunate that Fluehr-Lobban's review (2000) repeats mistakes concerning the history of anthropology as produced in Gregory and Sanjek (1994), Harris (1968), Williams (1996), and Stocking (1968). As I have shown in a recent work[2] these authors confused the efforts of plantation owners, slavers and others who profited from the idea of Africans being inferior with those scientists who were beginning to form the discipline of anthropology. Many of those cited were not anthropologists but biologists like Broca. Other examples are due to the confusion of the citation of writings of racists with the author's own works as when Williams cites Boas but it is a section where Boas really is using a quotation from someone else. This gets repeated due to the failing of those who quote Williams and Stocking to go back and check the original sources.

While the introductions (especially that by Fluehr-Lollan) are effective and bring the life of Firmin to light, what these lack is a comprehensive introduction that can place the history of the polygenist debate in the context of the development of biology as a science, and clarify the role of slavery and plantation owners in the struggle for anthropology. As it stands now, the contending anthropologies in the nineteenth century are conflated; it is as if someone wrote a history of climate science today without differentiating those who are paid by fossil fuel companies or have anti-science religious views and placed all into one camp and called them "climate scientists."

Our short travel through the history of Firmin and nineteenth-century racism was an example of the failure to see the world as reality due to privilege. Being unable to participate in the world fully, as mentioned above, as a human sometimes is due to organic disease or the result of trauma. This state was referred to in some examples from Oliver Sacks's work above and will be revisited in the next chapter. Being able to learn how to be a human is a process of enculturation and requires a human context as well as a normal human ability to learn. What happens to the learning process when either of these parts is not present in whole or part? This is our next subject of investigation.

NOTES

1. See Niccolo Caldararo, "Evolutionary aspects of disease avoidance: the role of disease in the development of complex society" (2012), https://papers.ssrn.com/sol3/papers.cfm?abstract_id=2001098, accessed January 1, 2017.
2. Niccolo Caldararo, *The Psychic Unity of Mankind: The Origins of Anthropology, the Anti-slavery Movement, Cultural Relativism and Man's "Unique" Nature* (Saarbrucken: Scholars' Press, 2014).

Chapter Eleven

Learning and "Hardwiring"

Even the closest of friends or lovers sometimes have disagreements. Humans often fail to understand each other because they speak different languages, and various cultures' depictions of humans and animals speaking with and understanding each other occur only in myth. Yet we have spent millions trying to teach primates to speak English. To Wittgenstein, there were more philosophical problems that led him to say, "If a lion could talk we would not understand him" (Wittgenstein, 1953). Steven Budiansky (1998) responds that if a lion could talk, he would not have a lion's brain, and therefore would not be a lion. Yet today, we create chimeric unions of mouse, pig and human tissue and transfer genes among them. So what, then, comes of "being a lion?"

The question, "What is learning?" has a long history of consideration. As habitats for animals disappear in the wild, our opportunities to study different species' learning in the natural world is diminished and ideas of "knowledge bases" of learning passed down from one generation to another is lost. If one considers the potential for the loss of human knowledge in just one generation, for example, in the destruction of the entire edifice of Native American learning by Europeans in the sixteenth century (Caldararo, 1994), it is apparent that the potential for its recreation is near zero. We are experiencing a flood of disease in children, some diagnosed as ADD/ADHD and some as autism. As we have changed our means of communication with computers this has had dramatic effects in learning and teaching. Are these diseases of the superorganism or its development?

Macphail (1982) makes a careful distinction between ideas of instinct and modifications of physiological states (habituation, adaptation, etc.) and learning, while Bonner (1980) asked if these do not represent a continuum. Denton et al. (1996), writing on basic mechanisms to maintain life processes, related to adaptation to changing conditions of life in simple as well as complex organisms including fish, birds and mammals, come to similar conclusions, that the difference between instinct (or "fixed" response patterns) and learning is more of a continuum in evolution as well as ontogeny, which corresponds to positions taken by Wilson (1975) and Gould (1977).

But the process of learning is often considered so easy that people are surprised when difficulties arise and children do not learn, or learn at rates considered to be abnormal. There are a number of conditions that limit learning, some of these are genetic in origin as in dyspraxia related to the FoxP2 gene (Geschwind and Rakic, 2013), others are disabilities in learning as in dyslexia, some are due to injuries including damage to language areas as in Broca's or Werneke's areas and related structures (Geschwind, 1970). One of the most challenging and controversial is that problem in learning termed ADD/ADHD. This disease also encompasses the problem of changes in consciousness due to problems in perception in development, yet changes in consciousness, self-image and reality result from a variety of factors like aging (e.g., dementias), diseases and trauma.

Oliver Sacks (1970) attempts to address the issue of a loss of function and its effect on the personality and nature of self-image and cognition (e.g., Korsakov's syndrome or anterograde amnesia). The outcomes are usually tragic, yet instructive. We have already considered the lost of identity and humanity produced by a racist society, but here the pathology is more diffuse, as Laing (1960) describes for the loss of self, the aftermath varies between not so much society and the individual in a false disability, but between child and parent and the nature of the interpretation the child makes of its own biology.

The issue is complicated by the structure of our own perceptions. Given that it is more than 40 years since Griffin (1976) made his criticisms of the behaviorist rejection of animal consciousness, it is amazing to read a scientist write that humans have "the ability to consider alternative strategies rather than blindly reacting to the events occurring around them" (Carpendale and Frayn, 2016). Use of terms like "aware" and "meaning" of actions used for human behavior but not for insects implies the authors know that insects think (or not). The authors claim that intelligence develops within social relations; of course, this is limited by social relations in economic panics or booms: the recent one in 2008 not only described a herding behavior but also the nature of magic and irrationality (Caldararo, 2009c, 2013). Given that attention is one of the most significant attributes of human behavior in that it can be communicated over time and space, it is curious how we have seen a condition that afflicts or distorts attention come to affect so many children. This significance of attention was discussed in relation to Graziano (2013) and his concept of attention schema theory. Is this condition a result of social complexity, and if so, how is it manifested and how will it affect the future of complexity?

Regarding the "illness" or loss of normality involved in ADD/ADHD, there seem to be two bodies of information that are being used to construct the idea of ADD/ADHD as a disease. One body of knowledge is derived from parents, teachers and clinical psychologists. This knowledge has been used to establish foundations for what to expect in a child who is performing poorly in school or to a parent's or other responsible adult's expectations of how the child should behave. From this foundation a set of normality guidelines has been fashioned and these have been used as baselines for clinical studies. No effort has been made to chart expectations of children's behavior over time and how ideas of normality have been conceived in, say, the past 200 years. Departures from these guidelines describe a child as abnormal and possibly pathological. Questions have arisen recently as to the independence of some of the clinical data from drug companies (Lenzer, 2008). Little work has been done to determine if there are cross-cultural adjuncts to these criteria of normality or pathology, though recent evidence is contradictory, for example, Brewis et al. (2000, 2002), Caldararo (2002a), and Jacobson (2002).

The second body of knowledge is derived from studies of the brain using MRI and other similar instrumentation, comparing children with clinically defined ADD/ADHD with controls. Also, this knowledge has been added to from studies of genetic markers and nucleotide sequences, especially copy number variants (CNVs) (Martin et al., 2014). Candidates include genes NDE1, LIS1 and DISC1 (Hennah and Porteous, 2009). Competing genes are related to dopamine, DRD4 and DAT1 (Volkow et al., 2010). Here, in both cases, we have no means of determining if the features seen in the MRI or the genetic information is the cause or the result of some exterior factor unknown to researchers. The major problem underlying this data is the fact that not all cases of ADD/ADHD demonstrate similar MRI features (while some controls without the diagnosis do have these features) and few with ADD/ADHD share the genetic variants while many controls have them but not the diagnosis. Difficulties with machine variations and with interpretations from small population samples remain to determine normality and development of abnormality (Sowell et al., 2004). A recent survey of CNVs (Conrad et al., 2010) concluded that, "The mere presence or absence of a variant in such control data sets is only partially informative, as the determination of pathogenicity of inherited CNVs is at present limited by the lack of information on their frequency and combination in apparently healthy individuals at a given age." Low correlation undermines association with clinical conditions (Elia et al., 2012). CNVs are the result of a diverse group of mutational mechanisms (Gu et al., 2008). Mutations appear in some cases, as in mitochondrial DNA, with a considerable degree of uniformity but not all of which are inherited but rather dependent on aging (Wang et al., 2001). The finding by

Conrad et al. (2010) that CNV breakpoints preferentially appear at recombination hot spots also demonstrates a pattern of causality. An understanding of the natural history of mitochondria and the nature of human diversity is central to considering the potential pathogenicity of such mutations and their inheritance (Caldararo and Guthrie, 1998).

The major question regarding CNVs centers on the use of the terms "associated" and "involved," "predispose" as opposed to "demonstrated." This is clearly shown in the published catalogues of diseases where CNVs have been generally found in frequencies of 1–3 percent of cases (Cook and Scherer, 2009; Elia et al., 2012). Can all the candidate genes be involved by association or do they all simply predispose the conditions of ADD/ADHD in different ways? Perhaps those genes involved in brain development have altered the reward pathways in ways that produce dysfunctional dopamine production or utilization. This could be interpreted from the fMRI data of some studies (Strohle et al., 2008).

One of the great advances in the treatment of disease in the nineteenth century was the development of Koch's Postulates that allowed practitioners to identify the causative agents of disease by the construction of diagnostic tools and techniques to describe the etiology of a disease. Koch understood the variation in the natural history of disease in the individual host given asymptomatic carriers and noninfection of some potential hosts after exposure. Molecular approaches to Koch's Postulates have also adapted our understanding of disease expression, latency and varied expression (Walker et al., 2006).

America is in the middle of a drug war that seems as impossible to withdraw from as Vietnam and Iraq and yet we see daily images of sports heroes, trainers and their doctors explaining the use of steroids. A survey of 200 hospitals across the nation by the National Association of Counties found that 73 percent reported an increase in meth related emergency room visits since the year 2000. On the other hand, we have never prescribed so many drugs for children, especially depressants. In just over the counter drugs alone research by Dr. David Rothner (2001) at The Cleveland Clinic found that 22 percent of 680 patients referred to the hospital's pediatric headache center were overusing nonprescription drugs. In a study published in *Drug and Alcohol Dependence* (Kroutil et al., 2006) the authors cite evidence that 1 in 11 children in America have been treated with a drug for ADD or ADHD. One has to ask why this contradictory situation has taken place. Culturally we would rather risk death by heart failure than allow people to treat pain with marijuana, given the Vioxx scandal. We use drugs like Ritalin for a diffuse group of diseases like ADD/ADHD and yet group them with syndromes like Tourette's. The variety of symptoms and potential links between them in the DSM-IV is reminiscent of the focal disease fiasco in the twentieth century (Anderson, 1996). Brewis and Schmidt (2003) have shown how a child's gender influences the way psychiatric symptoms are ascribed to them by some professionals. The criteria for diagnosis for ADD/ADHD seems to lack consistency and uniformity in application.

METHODS, THEORY, AND INTERPRETATION OF FINDINGS IN THE LITERATURE

Cultural Data

Interpretation of the first body of evidence about ADD/ADHD concerns goals of actors. Parents want to know how to understand and deal with behavior they find strange or uncomfortable, or disorderly. Explanations of the behavior may not be as important as how to change it. Many parents are willing to intervene in daily routines, TV exposure, diet, exercise, and so on. But often the call for change is constructed in an urgent context and this seems to affect both the relief sought by parents and the goals suggested by practitioners (Shaw et al., 2003). Constance Holden's report in *Science* (June 4, 2004) on John Warner's research in England (McCann et al., 2007) with food additives and ADHD children gives us reason to reflect on the massive drugging of children and the effects this has on development. In addition, we have a report by Christakis et al. (2004), who found that early television exposure is associated with attention problems by age seven. Zuckerman et al. (1987) found that children with persistent sleep problems are more likely to have behavioral problems. But all these bring up a number of points about the use of medication on children for behavioral problems, especially ADD and ADHD. Michael S. Jellinek's comments (2003) were chilling, as he notes that the economics of treatment has an effect on the epidemiological and clinical data. Anthropological research in cross-cultural context has often informed issues in medicine and psychology. Reference to this research is perhaps now useful in the present case with behavioral problems of children in the late twentieth and early twenty-first centuries.

As reported in the press (Waters, 2004) the Food and Drug Administration (FDA) rejected the recommendation of a medical reviewer that the agency should adopt a tough risk management strategy because the risk of suicide was so high among children taking antidepressants. The reviewer, Dr. Andrew Mosholder, an epidemiologist in the FDA's Office of Drug Safety, analyzed 22 clinical trials of 9 antidepressants and concluded that the drugs appeared to double the risk of suicidal behavior among children. Joan Ryan's March 23, 2004 article in the *San Francisco Chronicle* on antidepressants was also most disturbing. The relationship between the use of various drugs to control behavior and treat depression has reached epidemic proportions. Some physicians downplayed these results in the article, such as the opinions of Professor

Ponton and those of S.F. Suicide Prevention Director Eve Meyer, which are informed as they reflect on the symptoms of the cycle of drug treatment we have entered into in the last 60 years.

An annual analysis of drug use trends by Medco Health Solutions, Inc. found that a 49-percent rise in the use of ADD/ADHD drugs by children under five in the three years prior to 2004 contributed to a 23-percent increase in usage for all children. A survey in 2007 covering a five-year period from 2000 to 2005 found that 4.4 percent of children were treated, with a significant difference by gender and age (Castle et al., 2007). Figures compiled by the *Globe and Mail* in Canada showed that prescriptions for Ritalin and other amphetamine-like drugs for ADHD increased to 2.0 million in 2009, a jump of more than 55 percent in four years. Data on the CDC website, published on February 13, 2017, show a concentration of cases in the eastern United States with approximately 11 percent of children aged 4 to 17 years diagnosed as of 2011. The authors cited in the *Globe and Mail* study questioned whether parents were having doctors treat a disease or the behavior of "boyhood" (Abraham, 2010). In fact, spending on medicines for ADD, ADHD and other behavior disorders in children has now surpassed spending on drugs for asthma and infections in children. A paper by Mark S. Ansorge et al., published in *Science* (2004), reports on experiments with mice and Prozac. They found that treatment with Prozac early in development increased the appearance of depression and anxiety symptoms.

Some of the literature on the subject of ADD/ADHD uses terms in different ways with distinctly different meanings. One example is a recent article by Daniel T. Willingham (2004). Willingham uses the term "cross-cultural" with reference to work published by Szatmari et al. (1992) to state that ADD/ADHD is a syndrome found cross-culturally. The problem with this statement is that Szatmari et al. (1992) did not engage in a cross-cultural study. The data they cite all involved children with European backgrounds in the United States, Puerto Rico, West Germany, and Canada. Only a study in New Zealand included any number of children derived from non-Western cultural settings and the authors of this study (Anderson et al., 1987) clearly state these children (all Maori) were a small part of their sample and underrepresented. The language used to discuss the syndrome and the means to treat the symptoms is less scientific and more journalistic in nature. There has been significant disagreement on this point (see Dwivedi and Banhatti, 2005; Faraone et al., 2003; Hinshaw et al., 2011; Moon, 2012; Timimi and Taylor, 2003).

Scientifically, this syndrome is quite controversial, and many deny its existence. In an article in *American Anthropologist* (Caldararo, 2002a), I argued that these syndromes are culture-bound effects, but could they be also technological aftereffects? Perhaps in the current milieu we can make some assumptions to explain this. One, where parents have little time to provide the training regimes that result in the kind of childhood behavior they feel is appropriate, or two, where they are incapable of such training. Or three, our society is demanding less order in the home and schools, resulting in children who are less able to respond with the kinds of normative behavior expected of them. There is yet another possibility, that of the effects of computers and computer screens and programs on the cognitive development of children.

Perhaps, though, the variety of behavior we see in children today is no different than that of 30 years ago, but we are expecting order, or it may be that hyperactivity is the result of exposure to too much TV or a change in the kinds of programming typical of the past 60 years of broadcast TV. Such developmental stress may be responsible for the developmental trajectories found by Castellanos et al. (2002). Whatever the cause, we find several recent articles by Ken Jacobson (University of Massachusetts-Amherst) commenting on his cross-cultural research on ADHD, supporting my position that ADHD and ADD are culture-bound syndromes (Jacobson, 2002).

As the behavioral data and the diagnostic criteria seem subjective, we cannot be sure. Yet some reports clearly point to a specific syndrome though with factors that may vary from society to society. The ADHD Institute (http://adhd-institute.com/burden-of-adhd/epidemiology/) reports a wide variation in the distribution and frequency of cases citing a number of studies. Some (Kuruppuarachchi, & Wijeratne, 2004) argue the syndrome is limited to developed countries, while this is contradicted by others (Chinawa et al., 2014).

Jacobson has shown that ADHD-like behaviors are found widespread among children labeled as "normal" in different societies. And, most importantly, he found significant differences between English and American teacher standards with respect to what constituted "appropriate" classroom behaviors. Thus, there is some basis for arguing that children are not behaving differently, but are being perceived as different. What is strikingly unsettling, however, is the speed with which drugging becomes the answer, both in the case of supposed ADHD/ADD and depression. While we are learning much about the effects of nutrition, exercise and different foods on children and the period of adolescence, it is alarming that we are so willing to drug children with substances that have not been studied for their effects on children over any considerable length of time. While rat, mouse and nonhuman primate studies may provide general ideas of the effects of such drugs on maturation, the human infant and child undergoes a unique process of encephalization. We might pause to consider if the social cost of drugging an unruly future Einstein is worth the silence.

INSTRUMENTAL AND GENETIC DATA

In many cases, MRI imaging of the brains of children diagnosed with ADD/ADHD fail to show the variation of supposed brain abnormalities compared to controls (Sowell et al., 2003). The two groups are simply referred to as means in many studies. We do not know if there is an overlap between the groups and how much the results are the result of independent variation. Individuals with ADD/ADHD are shown as smaller and morphological differences may be the result of nutrition or growth differences. Such grouping of individuals occurs in other MRI studies where individual variation is ignored (Thompson et al., 2004). In the Thompson et al. study, the study group consisted of methamphetamine users who were also HIV positive. No reference was made to abnormalities compared to controls that might have resulted from HIV infection. Certainly previous drug use may explain the described changes in the Sowell et al. (2003), ADD/ADHD population (though Castellanos et al. seem to contradict this idea), but also behaviorally induced trauma or diet might also be implicated. On the other hand, sleep problems induced by behavioral stress including TV (and computer and smart phone screens) could explain the changes in brain structure. A study of the effects of computer screen blue light by Harvard University researchers (Harvard Medical School, 2012) found significant problems in sleep and health outcomes. More recent fMRI use (Adisetiyo et al., 2014) indicates that differences in brain iron levels between use of anti-ADD/ADHD drugs and untreated controls may be diagnostic.

We should be cautious about the idea of gene-disease association, as recent research has shown (Lucentini, 2004). Reproducible gene-disease associations are few. The D3 receptor gene with its supposed association with schizophrenia is a good example (Jonsson, 2004). While the specific criteria for causal disease association have been questioned (Fredricks and Relman, 1996), with a greater emphasis on genetic variation and susceptibility of individuals and variability in disease state, especially with diseases that result from viral infections and are associated with long latent periods (Caldararo, 1996), the complexity of gene involvement has been demonstrated in a growing body of literature on single nucleotide polymorphisms (Bergman et al., 2010) and other factors including histones (Caldararo, 2002b) and disease expression and progression (Caldararo, 2003a).

The interaction of influences on ontogenetic development, maternal status and diet can have significant effects on some of the genetic data. A study by Eisenberg et al. (2008) showed that DRD4 differences in Body Mass Index were due primarily to differences in fat-free body mass. The relationship between differences in dopamine receptor pathways in different populations needs more study to understand any etiology of disease as other factors may be involved in specific populations that have been tested. For example, the populations of ADHD subjects and controls constructed by Volkow et al. (2010) possessed fewer than 5 percent smokers, certainly an unrepresentative group.

Studies in cross-cultural contexts of children and nutrition (see Barry Bogin, 1999, 2001) have produced a vast amount of new information, little of which has been considered in the use of drugs to control behavior. It may be that we are producing abnormal growth and behavior by drugging children which later appears as needing more drugging to treat depression and other problems. Ilina Singh's recent article, "Doing their jobs: mothering with Ritalin in a culture of mother-blame" (2004), delineates the cultural foundations for drugging children.

CONCLUSION

What is equally amazing when one reads of the extent of the use of drugs on children and adolescents, is that while we are relatively unsure how these drugs will affect development, we deny young adults access to drugs (mostly the alkaloids) which have been used for millennia for the purpose of effecting the transition from childhood to adulthood. See the writings of Peter Furst (*Flesh of the Gods: The Ritual Use of Hallucinogens*, 1972 and *Hallucinogens and Culture,* 1977) and Michael Harner (*Hallucinogens and Shamanism,* 1973). Our society punishes doctors on the one hand, if they medicate people in chronic pain to the point where they can live without pain, and yet we encourage our youth to drink alcohol to dull their senses. Our culture seems to prefer to have our children consume substances like Ritalin and sit stupefied in the classroom like robots, rather than take a natural drug like mescaline and explore the spiritual worlds of their ancestors. One has to wonder if there is a "war on drugs" given these circumstances. The Mayo Clinic (2013) reports that 70 percent of all Americans are using prescription drugs.[1] Is this the inexorable "way" of the superorganism, and do we believe this is a positive path? Technology seems today to do no wrong. As the Grateful Dead might have put it, what a strange trip we all are on!

NOTE

1. See "Nearly 7 in 10 Americans Take Prescription Drugs, Mayo Clinic, Olmsted Medical Center Find," *Mayo Clinic News Network*, http://newsnetwork.mayoclinic.org/discussion/nearly-7-in-10-americans-take-prescription-drugs-mayo-clinic-olmsted-medical-center-find/, accessed March 17, 2017.

Chapter Twelve

The Housing Crisis and Homelessness

> One of the most interesting results of this investigation is the discovery of a class of persons unattached to any gens, "outcasts from the bond of kinship." (Lewis Henry Morgan, 1868)

Homelessness has many factors. The nature of homes or places of residence versus camps can be defined historically and culturally for our species. Many animals produce structures to live in, others excavate depressions, many live in caves or inhabit structures built by other animals and discarded. In many cases, animals simply find limited shelter in temporary settings—to avoid weather conditions or to rest—building nests to sleep in or utilizing existing foliage. Humans have, over the past two million years, behaved in all these ways. Structures are relatively new, appearing in the last half million years, while some peoples build only temporary windbreaks. The nature of homelessness is a modern classification, though some sources from the Greek, Roman, Indian, and Chinese societies have commented on its varieties. Most modern discussions of homelessness address it as a phenomenon of modern industrial society based on inequalities inherent in modern capitalism. These include relationships homeless individuals create with residents, including economic behavior. This chapter attempts to construct a different focus on how space is used by the homeless, how this is viewed by residents and those working only in the day, and also looks at the kinds of relationships that develop between these groups over time.

If humans lacked "homes" 10,000 years ago, why must we now have them? People without homes can still be part of the superorganism today: they can own smart phones, have bank accounts via the Internet, and engage in commerce (Tucker, 2014). Being a nomad was the human experience until very recently. From the huts at Terra Amata, dated to about 400,000 BP, to tents, cardboard boxes, and autos, humans have mostly lived a mobile life. Do we lose some quality of being human without a home, or a single location to call a home, or is this a loss of our humanity? Why do we find increasingly forced housing on people across the globe? If the elements of the transition to agriculture created sedentism, then a post-industrial society should find homelessness more compatible as people today are nomads following jobs from place to place. New ideas of work and living spaces are increasingly described as common with wall-less offices and cooperative work areas. Millions today live in illegal shantytowns much like those of the Industrial Revolution. Why are the homeless such a problem? One might think that those who desire homes must be more of a threat to the superorganism, they would be less plastic, less free, and more rooted.

BACKGROUND

There is a worldwide housing crisis, in developed as well as developing countries. The credit crisis that began in 2008 is often blamed on a crisis in the housing industry, but this crisis in the industry has its roots in problems in housing America's population. How much housing is needed? What kind of housing is to be built and how big should these houses be? And the central, most significant question: How expensive or "affordable" should the housing become?

It has become clear to many architects, planners, and social scientists in the past 50 years, after comprehensive studies of development practices in the United States and abroad, that planning practices resulting in high-density inner cities and sprawling suburbs produced not only economic and social problems like congestion and pollution, but also disease, both physical and psychological. In a classical study published in 1967 of the key factors regulating animal populations, D.G. Harcourt and E.J. Leroux found that most factors, like disease, parasitism, underdevelopment, and so on were density dependent. This means that the denser the population, the less healthy the environment. In a comprehensive study of cities the world over, the United Nations' Habitat study *State of the World's Cities*, published in 2006, demonstrated that people living in cities were not better off, and were sometimes worse off, than those living in rural areas when compared to a plethora of health criteria. Mike Davis' 2006 book *Planet of Slums* argues, from an analysis of the expanding cities and suburbs, that they leave little space for social interaction and produce crowded, poorly serviced living areas ripe for violence and crime.

This encompasses the Modern Movement in architecture (Blake, 1976) though the production of multi-story, densely designed tenements was common by the late nineteenth century in America, and earlier in England. It is reasonable to say that density is relative in human history given the contrast in lifestyle between a nomadic people and the sedentary life of early urban societies (Adams, 1966, 1981).

However, it is clear that when the human adventure began some seven million years ago with the partial bipedalism of *Sahelanthropus*, hominids lived in the open. One might say that hominids were "homeless" in the sense that the first evidence we have of build habitations of hominids is dated to about 400,000 years ago (Edwards and Clinnick, 1980). This would be technically correct if we equal not having a built home with being homeless, yet hominids, like most mammals have had core areas in established ranges that can be considered their "home." These are areas, as in baboons or chimps or gibbons as well as others, where they can most often be found and preferentially exploit food resources, safety areas, and water (Dolhinow, 1965).

The Terra Amata structures were flimsy windbreaks made of branches and interwoven to produce an effective habitation (Villa, 1983). They appear to have been used only temporarily and then abandoned, possibly to be reused later, as there is evidence they were rebuilt in the same area over and over again. Still, we might be correct in saying that from seven million years ago to 400,000 years ago, hominids largely lived without "built homes." After that I think we would be correct, given the few examples of such structures and there temporary use, that most hominids lived like today's homeless, creating temporary structures and moving on to other locations. Obviously, in the past, this was associated with a complex particular and localized culture that occupied areas for their own use. Today's homeless do not have such control over sites they use (Bourgeois, 2009) for a number of reasons.

Should we then trace the appearance of permanent structures that hominids made, we would find our evidence leads us to a period after 20,000 years ago, although a structure found in Germany at Bruniquel that was built by Neanderthals appears quite substantial, and though built in a cave area, it could be considered a permanent structure (Balter, 1996). Spivak and Nadel (2016) argue for a date of about 23,000 BP based on Radiocarbon dating at Ohalo II in Palestine where remnants of dwellings made of sticks and mud were found. Earliest evidence of substantial structures is reported from the Natufian-PPNB culture (Eckmeier, 2012; Steadman, 2004) and is dated after 15,000 BP. Thus, we can regard the establishment of housing and the beginning of homelessness to this period. The way people are living today, as we will detail below, with no secure job, no affordable home, and little in the way of a common place associated with a community, is unique in human history. It is unlike the nomadic life of hunter-gatherers or like the communities of sedentary societies over the past 15,000 years. In fact, the twentieth century, perhaps as no other time in human history, has seen a massive migration, but in some cases, this migration has been within countries into refugee camps but still some distance away from their original homes. These migrations are made up of crowds of those fleeing danger, violence, or war, and these uprooted people are most often stripped of their property as no other group of migrating people in history has been. In other examples, as in the refugee crisis following the wars in Iraq, Syria, and Afghanistan, hundreds of thousands of people have traveled thousands of miles to find safety, food, and new lives in Europe (Maloney and Korinek, 2011).

The migration of Europeans after 1492 was substantial and forced the migration of millions of Africans to the New World as well, and there have been internal migrations, some voluntary as in that which followed the Dust Bowl and Great Depression in the United States, and some involuntary as those within the Soviet Union after 1920. But the creation of refugee camps and homeless encampments in the twentieth century were temporary, until the late twentieth century and the early twenty-first century where refugee camps and homeless, squatter camps became semi-permanent arrangements (Neuwirth, 2006).

A SPACE PROBLEM, NOT A HOUSING PROBLEM

The South of Market area in San Francisco has a long history of occupation by immigrants and workers amid scattered factories. The area was considerably depressed in the 1970s as many factories moved to lower-rent locations in other parts of the Bay Area and as demographic shifts due to S. F. Redevelopment Agency activity in the eastern sector of the area resulted in the demolition of single-room occupancy buildings and factories. A general plan for the area emphasized clearance and redevelopment (Hartmann and Averbach, 1974). Since the 1970s, redevelopment has taken place farther east and south, especially concentrated in China Basin and the Third Street Corridor.

Some urban ethnographies are situated within the economic activities of the focus population as in the case of Mathews' (2011) study of Chungking Mansions and Bourgois' (2003) study of Puerto Rican drug merchants. Peripherally, in both, we learn how people use housing as a temporary platform for economic survival. From the experience of Mathews, merchants from Africa and India use the premises of Chungking Mansions as a multinational market, and the smaller and cheaper the space, the better to carry out their long-distance trade. For Bourgois, living spaces are temporary locations negotiated within fragile adult male and female relationships, dramatically affected by changing patterns in New York's economy and racist concepts concerning minorities as well as female roles in the economy. Children learn not only the temporary nature of housing—moving between nuclear family arrangements to grandparents—but also the temporary nature of male status in homes.

In the present study of this report, we focus on the out-of-doors space and how home and recognition of "home" are constructed between homeless individuals, local residents, and workers. It began in 1980 when the author commenced living in the South of Market area on Fifth and Folsom Streets. Then, the presence of homeless individuals was minor and relatively unnoticed by most residents. After the 1989 earthquake, many buildings in the area became uninhabitable and attracted the attention of homeless and semi-homeless individuals (drug users and suppliers, artists and sexual adventurers). The latter category was partly homeless by choice as they used abandoned buildings as temporary locations for their activities and had, or claimed, other residences. This research project changed in nature when the author moved out of the area in 1992, but occasional research trips continued into 2010.

TYPES OF HOMES

Since this section addresses homelessness, it seems appropriate to attempt a definition or at least a background context for what homelessness is, and perhaps some sort of a sketch of the urban environment in which it is most often described as a problem.

Studies of how people live and organize their living spaces are not new. Lewis Henry Morgan wrote his comprehensive study of Native Americans, *Houses and House-Life of the American Aborigines*, in 1881, and Edward T. Hall's *The Hidden Dimension* was published in 1956. The latter focused on the necessary optimum space for people to live and work effectively, with the least detrimental outcomes from a cross-cultural perspective. It should be recognized that in some contexts, homelessness and poverty are associated with the stranger. Abi'l-Hadid notes this regarding the poetry of Halaf al-Ahmar (d. 796 CE): "Don't think that a stranger is the one who is far away! Rather, a stranger is the one who has little or nothing (al-muqill)" (Ibn Abi'l-Hadid, 1963–1964).

A wave of criticism of the theories of modern housing has appeared since the 1960s where large housing projects were associated in the post–World War II period with crime and poverty. In Europe, the projects of architect Le Corbusier have been blamed for the lack of social life that contributed to the recent riots in France and those in the 1980s and 1990s in England (Aspden, 2006). A transformation of housing took place under a number of architects, however, the "Master Builders" of the twentieth century as Peter Blake calls them, including Mies van der Rohe and Frank Lloyd Wright (Blake, 1976), were responsible.

American architect and critic Charles Jencks (2000, 2007) has summarized problems of the Modernist Movement in concrete blocks of housing and central city office buildings in a number of works. Among the first coherent criticisms of this planning were Jane Jacobs's *The Economy of Cities* (1969) and Robert Caro's damning biography, *The Power Broker: Robert Moses and the Decline of New York*. Their analyses describe how building policies in the 1950s and 1960s contributed to New York's economic and social problems of the 1970s and 1980s.

In recent years, efforts by architects and developers to rebuild such structures and rehabilitate old factories by designing social spaces, and common living areas and vistas have produced a new perspective in the use of space for living and working. This work revisits, without reference, the earlier work of social scientists like Redfield and Singer (1954) and Robert Ezra Park and his associates (1952 and 1967) who examined the role of the city in history in terms of the satisfactions provided for the daily life of individuals and in the development or decline and transformation of culture.

Following several decades of redevelopment of aging city centers and the trends of the 1980s and 1990s of rebuilding city cores as lifestyle work landscapes ("gentrification" to some critics), advertising agencies have undertaken surveys to characterize the idealized and modal living spaces of different national populations given different incomes. Emily Backus (2009) has outlined some of these efforts, called "Cross-Cultural Consumer Characterization," conceptual frameworks as designed by Young and Rubicam, while another firm, Makno Consulting, has produced a "Housing Evolution Study," identifying seven types of home. It is less an evolutionary study than an attitudinal one. Attempts to address differences in the need for housing, how people of different cultures and access to wealth define their requirements and perceptions of adequate housing remain difficult to compare. Fossum (1965) reevaluated studies in the twentieth century that found efforts to associate types of housing with social or health conditions (crime, disease, education, etc.) were undermined by various definitions of values across cultures, caste and class. As Neuwirth (2006) notes, definitions of slum housing have changed dramatically in just a few decades, and the same locations that were once considered dangerous, characterized by poverty, and illegal have recently been attracting upper-class residents. Adequate housing in the United States is quite different from that in Kenya or Indonesia; the type of house that is acceptable or desired and associated with a healthy environment varies considerably. How are these facts to be reconciled in a global environment?

Out of this scenario we find, in the past decade, a growing international response aimed to identify the kinds of qualities in living environments that produce the most positive social interactions and health outcomes. This movement, a combination of work by architects, developers, builders, and social scientists, has loosely been called the Home Zone or Shared Space planning approach. One of its conclusions is the need for open space that is practical for use by a community and designed and maintained by localities. Dutch engineer Hans Monderman is often referred to as the "father" of the Shared Space idea (Vanderbilt, 2008) and he acknowledges that he derived much of his approach from studying traditional and ancient villages. The question of the nature of the pre-industrial village has resulted in disagreements between historians, social scientists, and archaeologists. One summary of the issues involved appears in Sjoberg (1960). Other sources of cultural living patterns for home organization have become popular as in the case of Feng Shui and Vastu shastra (Lawlor, 1994).

In the April 28–29, 2012, weekend issue of the *Financial Times*, in the "House and Home" section, there is a study on second, third, fourth, and more houses of the wealthy and the effects such housing has on cities. (This "buy to leave" strategy was revisited in a 2016 article by Graham Norwood.) Generally, the effects are negative. Cities get less in sales tax, have less sales in areas, less foot traffic, neighborhoods become ghost towns, and character is lost. This is consonant with the findings of Saskia Sassen of Columbia University in her book *Cities in the World Economy* (2011). New "ghost apartments" in luxury buildings have cropped up across the globe and consume materials, labor, and investment that otherwise could produce affordable and safe housing (Johnson, 2015).

I have been using Marin County in California as a laboratory for studying the problems in the creation of low-cost housing. While much of the country suffered after 2008 from the congested foreclosure situation relating to tranches of mortgage loans sold as asset-backed securities in the past decade, in Marin, the competition for land by luxury home builders and resistant existing homeowners has made low-cost developments very difficult. In this, we find that absentee tenants now own many housing units in West Marin that were used for agricultural workers. These vacation homes are making existing housing more expensive.

A 2011 study by the National Home Builders' Association found that 5 percent of America's housing stock is held as second homes.[1] Nationwide, the 1990 Census found that about 3.1 percent of the housing stock of the country was held as second homes,[2] with California registering about 2 percent of its housing stock as second home status. This figure has been relatively stable since 1950, rising about 1 percent from 1940 to 1950. There is evidence that second home sales soared from 1990 to 2011.

However, do we know how many homes in West Marin are second homes or vacation homes? Some are simply investments. The 2010 Census tells us that in some towns, like Ross, almost 10 percent of the housing units are vacant, while in the county as a whole, the rate is 7.2 percent. While all these units may not be second homes, the fact that so many homes are not being used for dwellings is certainly a waste.

Local governments are being pressed by the Association of Bay Area Governments, an agency that has a State mandate to produce housing target numbers, which are spurred by new State laws to build more housing. It seems contradictory that so many units are unused. This is worse if the real-use figures should be a combination of vacant homes—under construction, damaged or held off the market for investment purposes—and second and vacation homes. If this is the case, then the total underused housing is between 10 and 12 percent for California, as a conservative estimate, because the data also show that perhaps as many as 5 percent of condos and townhouses are held off the market as investments.

CITIES AND DENSITY

Some books have appeared that focus on single issues of qualities of cities, like Joel Kotkin's (2005) emphasis on security. This ignores what Redfield and Singer (1954) note as the two most important indications of cities: order and diversity. Given the history of cities, they argue that we must distinguish between those of homogeneous folk societies and those of heterogeneous centers of social and technological change and transformation, where people of different political and cultural groupings come together for industry and commerce. Some students of cities like Thomas Sieverts (2003), have argued that modern cities are growing without the traditional design of European "medieval" centers, but as multi-nodal or fractal units. The issue of cost of high-density versus low-density housing, or sprawl as opposed to urban centers has been found to reflect an actual infrastructure cost of low-density over high-density of 22 times more. Services for dispersed housing are also more costly as 1 house per acre or 640 per square mile versus 19,000 units per square mile in an urban core (Danielson and Lang, 1998; Leinberger, 2007).

The problem of space within a dwelling and its organization and use has also been related to the arrangement of houses to village or town structure, or the settlement pattern. Some attempts were made in the last century to study the use of space in a settlement and the relation house size had to settlement size and population, as in the work of Naroll (1962). The shape of building, especially size in relation to landscape features and horizon, has been investigated in relation to cultural values and power where Deyan Sudjic (2005) argues that power and the concentration of wealth drive the construction of monumental building. I have elsewhere concluded from a cross-cultural and historical analysis that wealth concentration was an important factor, but so was the emphasis and acceptance of the use of resources to express prestige in monumental buildings (Caldararo, 2004a).

The question of the use of space in a dwelling and in a city brings up another consideration regarding the use of resources for the production of housing. How much "house" is needed for each person? We do not seem to have a housing crisis in America; we have a space crisis. This is made clear every time there is a recession or a significant reduction in employment, as after the 1987 stock market collapse or the dot-com collapse or, now, with the current credit crisis and subprime debacle. At these times, people move in with friends or rent rooms in houses with strangers or family. We then hear of a crisis in the housing market due to oversupply. When times are good, people want more space and not only do we find a tight housing market but also pressure on small units on large parcels to be renovated and enlarged or replaced with larger structures (Local Government Association, 2009; Makin, 2006). One finding by Naroll (1962) was that as populations grew, so did the size of the average dwelling unit in a sample of 18 societies representing a number of Native American and Asian cultures (see his table 1 and figure 1). However, the nature of population pressure and inequality has resulted in modern conditions where living area is collapsing compared with these traditional society figures. We can compare China's 2004 figure for urban per capita living space at 25 square meters (269.1 square feet) (Xinhua, 2005), while for India, given 2006 urban data yields an average of 117 square feet (Thakur, 2008). Average Indian rural per capita living space is 65 square feet, about the size of the average American jail cell. India's urban space average has increased slightly since a 1953–1954 survey found an average of just less than 100 square feet (Thakur, 2008). But the central issue also is housing size and condition, as documented in the United Nations study for 2006.

It is our appetite for size that is causing the crisis in housing. We see it every day around us where small one- and two-bedroom, affordable houses in every town and city are being enlarged or destroyed to make way for bigger homes. Buildings where people raised a family of two or three children in a thousand square feet are giving way to homes for two people—if it is not a second home for a single family—of three or five thousand square feet. Big houses need more low-paid workers to act as nannies, gardeners, and so on, causing more transit demands for workers. The problem relates to the average size of living units since estimates of slum growth (Jargowsky, 2015) as a factor of urbanization versus the rural housing stock now being abandoned lack data, but recent studies in the United Kingdom found that as much as 10 percent of all construction materials brought to a building site are never used and that the waste produced from building was three times as great as all the household waste in the United Kingdom (DEFRA, 2006; Harvey, 2006).

Some research into traditional housing is being addressed by public entities (Duffles, 2006; Pollock, 2006) but more work needs to be done to develop sustainable housing concepts for transitional housing demanded by expanding modern urban centers that are creating slums. Current ideas for "eco-houses" seem appropriate for developed countries, but do not seem likely in the developing world, while proponents of some designs and materials such as *Building with Cob* by Adam Weismann and Katy Brice may have useful application in certain contexts, though self-made squatter housing seems more likely (Neuwirth, 2006).

Gentrification is not the only pressure on low-cost housing; the housing stock is constantly undergoing modification by rebuilds and additions. These are the least productive uses of resources. Housing is fashion for many people, not just in the middle class desire to possess a status home of a certain size, but in the inclusion in a house of certain amenities from mud rooms to dens and media rooms. Many traditional people used housing to display prestige, but today, we have reached a frenzy of building that is destructive in the use of land and resources. As the study by Nathaniel S. Keith (1973) demonstrated, we have duplicated the housing bubble of the 1920s that resulted in the crisis defined in 1937 where one third of America's families were found to be ill housed. There needs to be an international dialogue on housing, and anthropologists can contribute considerably to this project. What is an affordable house? We hear more discussion of "green housing" than low-cost housing today as if a green million-dollar house was affordable for America's median income family. However, 27 percent of all households in America are single people and the median house is over 2,000 square feet. Heating and cooling such a large space for one person makes even a "green" house expensive and unsustainable as a world model (Susanka, 2006).

Perhaps we do not need more housing, just more efficient use of existing housing and land. Many people seem to believe that more density is better, and this is true of our local builders and real estate lobby. They push "green building" as if it were an answer to more building, but as Jane Powell noted in her article in the *S.F. Chronicle Magazine* (2007), this can hardly be the answer, since demolishing 10,000 square feet of old buildings wipes out the environmental benefit of recycling 2,688,000 aluminum cans. No matter how many "green" products are used to build a house, the destruction of our small housing stock is rapidly making our housing inventory more expensive.

Studies have found that on the one hand, animals benefit from associating together but that at a certain point, which varies with the specific ecology and species, the denser the population, the less healthy the environment for individuals and groups (Allee, 1938, 1951). This contrasts with aspects of environmental enrichment experiments of Diamond and associates, where food and other conditions are not factors (Diamond, 1988). The March 31,

2007 issue of *The Economist* documents the catastrophic effects of migration to the cities in Africa and the economic and health problems it has created, while Moyo (2009) and Lancaster (1999) argue that foreign aid and charities support the growth of slums in African cities and promote depopulation of the rural villages. It seems obvious that there are too many people moving into cities in the developing world for those civil entities to cope adequately and that building homes in developed nations is utilizing too much energy. Neither case is sustainable. Paul Kennedy's tragic predictions in his 1993 book *Preparing for the 21st Century* are upon us.

THE CHARACTER OF HOMELESSNESS

Forms of housing and homelessness were discussed in numerous articles in the American Anthropology Association's newsletter *AN* in December 2008. Ways of living and of ordering people in different societies and in different economic systems has been an important issue in the popular media recently, but strangely, perhaps, without the input of anthropologists.

Margaret Mead in her book *Continuities in Cultural Evolution* (Mead, 1999) emphasized the fact that housing often has encoded in its design and construction, as well as the landscape it is set into, instructions for life and guidelines for behavior. On the one hand, in developed countries, these ancient patterns are being utilized in the design of the most expensive housing and imported from aboriginal contexts (Lawler, 1994). However, articles by Zeiderman (2008) and Greenbaum et al. (2008) seem to me most compelling at the moment in terms of displacement and resettlement. We should be more concerned with how people respond to changes in housing. Plutarch notes, concerning the Cyrenians in the first century BCE, that "nothing is more intractable than man when in felicity, nor anything more docile, when he has been reduced and humbled by fortune" (Plutarch, 1910). My experience with people being relocated from various areas of San Francisco by its Redevelopment Agency in the 1960s and 1970s taught me the truth of this observation and paralleled the conditions Max Weber (1904–1905) noted in the mass relocations in Europe in the nineteenth century. Nevertheless, we must consider carefully the types of choices people make about housing or to refuse housing. Just as the suppression of neighborhoods in nineteenth-century Paris and wide boulevards created cities that were easier to control by authorities, so today's cities are monuments for transient life with a workforce without job security, social integration (e.g., lack of rent control of kinship solidarity), and an atomized population without coherence or roots. Increasingly, people are periodically homeless, transhumant between temporary living arrangements and jobs.

In human history, there have always been homeless people. In some cases and in particular cultures and times in history, they have been known as sages, like Confucius and many thinkers and teachers of his time in China. A similar case can be seen in India where the sadhus and sramana, or holy men, are found wandering in the forests and towns (Eskildsen, 1998; Hartsuiker, 1993; Rousseau, 1978). A long history of secular and holy ascetics is known from the Mediterranean and a number of saints acquired their daily bread by begging. Travelers have often found refuge in towns and cities, in temples or parks (Anderson, 1961; Weaver and White, 1972). But homelessness and the homeless are generally regarded in economic terms today, and tolerance of the homeless can be measured in both cultural and economic terms throughout history. This is also reflected in the social history of terms used to characterize the homeless. In parts of the United States, and especially in the nineteenth and early twentieth centuries, homeless or transient workers were often called "hobos" or "tramps" or "vagrants," depending on their availability for work (Bruns, 1980; Higbie, 2003) or local concepts of criminality (Cook, 1979).

People without established homes who have an indigenous history are often called nomads or pastoralists depending on their technology and ethnic origins. They may also be classed as hunters and gatherers. However, economic change can upset established long-term relationships and produce homeless people from an underclass as in England from the sixteenth to nineteenth centuries (Beier, 1985; Polanyi, 1944).

Those who inhabit urban environments as a general routine with a certain modern cultural history are variously called hobos, tramps, or vagrants. In European history, the relations between seasonal workers of the villages and various towns defined a certain kind of temporary settlement of an economic nature. The forms of temporary urban settlement, or homelessness, caught the attention of many sociologists and historians of the nineteenth century, including Max Weber (1921), as temporary residence has had a role in economic development and urbanization. Foucault (1977) describes the transformation of traditional vagabondage into criminality during the eighteenth century as the labor needs of industrializing Europe increased.

Robert Ezra Park was a Chicago sociologist who was one of the first to study the mentality of domain of the homeless (Park, 1952) following the work of Roderick D. McKenzie (1933), who published on the ecology of communities in the 1930s and 1940s. McKenzie and Park had witnessed the dynamic mobility of Americans during the Depression and World War II and were interested in how quickly people adapted to changes of residence and homelessness (Park et al., 1967). What is most interesting in their work, however, and in regard to contemporary city and town's current problem with the homeless, is the idea of domain. In the 1960s, Thomas Weaver (Weaver and Douglas, 1972) applied ethnographic techniques to the intensive study of the homeless in the Seattle area. The result of his work was a classification of perceptual areas, or domains, that people construct about themselves, their residence, and the environment. Weaver found that people communicated to each other in how they used space and, when there was no agreement, then conflict took place.

Weaver found that people who are homeless are often temporarily so, and regard the state as one that is of a desired or necessary short duration. Often these were alcoholics who abandoned their homes to "binge" or to escape the confinements of home and job routines. People in neighborhoods, and city and town authorities regard such people with tolerance depending on their cultural traditions, religious background, education, and the standing of the individuals in the community (Asander, 1980; Guzewicz, 2000). Such "temporary" homelessness then serves a function to allow some people to maintain a degree of resilience in their relationships, although there are definite differences in homelessness, its relation to life history, its duration, and form between men and women (Passaro, 1996).

The domain of homelessness is then one that exists within an extended home and status in the community. The community tolerates the behavior as it is seen and accepted as an extension of the functional life of individuals. Forms of economic homelessness have been seen by some scientists as an increasing aspect of modern life where people are often one paycheck from losing their residence and reflect a lack of stability in American economic conditions and social welfare (Barak, 1991; Jencks, 1994). In 2008, the head of San Francisco's homeless program, Dariush Kayhan, reported (*San Francisco Chronicle,* July 17, 2008) the results of a citywide study that found that San Francisco's homeless conform to Weaver's model of patterned homelessness, showing a continued pattern over time. It would seem that San Francisco has succeeded in reducing some forms of homelessness, as well as overall numbers of habitual forms, as the result of the City's "Care not Cash" program where cash allocations in the form of welfare payments of general assistance checks were replaced by individual provision for shelter and supplies. This program's success contradicts World Bank reports that cash allocations to the poor are a more effective means of providing assistance that leads to sustained incomes out of poverty (see an article in *Financial Times* by Alan Beattie, Wednesday, August 6, 2008). San Francisco's results do parallel those in New York reported in March of 2009 (NYCDHS, 2009). However, the methods used to count the homeless are varied and under some criticism (Allday, 2009; Harrell, 2009; Marcus, 2006; O'Connor, 2009), and following the economic crisis of 2008 we saw counts go up. Data provided by a *San Francisco Chronicle* study indicate, however, that by June of 2014 there had been little change in the number of homeless individuals on the streets of the city and that about 19,500 people had been moved off the streets into various forms of lodging (Knight, 2014). While Noy (2009) argues from his study of framing theory that various actors in the city failed to act in cooperation and thus wasted efforts and resources, it is also possible that the new numbers represent the massive relocations and evictions of low-income people from housing due to significant rent increases and changes in properties from rentals to tenancies-in-common or owner move-in projects allowed under the state's Ellis Act (Sabatini, 2014).

Arguments have also been made that many homeless are veterans, former patients with continued psychological pathology, and victims of failed health-care programs. Philippe Bourgois and Jeff Schonberg (2009) have addressed these issues in a comprehensive analysis as well as the tendency to criminalize homelessness. In some cases, as in the research I undertook in San Francisco, the homeless provide services for residences over many years, watching homes when residents are at work, watching cars at night, or keeping tabs on neighbors, and in some cases, watching the activities of police.

Many communities have acted to eliminate homelessness of all these types, and some, like New York, have been rather successful. Usually, these efforts are temporary as municipalities often cannot afford to continue to direct resources to prevent homelessness, and campaigns to do so are usually politically motivated and short-lived. It seems that the first problem to overcome is to address the attitude that homelessness is an acceptable alternative lifestyle that people have the freedom to choose. This is a cultural and legal issue. There were towns in history without the homeless (Kenyatta, 1938), but these are unique cultural and historical situations. To eliminate homelessness requires the provision of resources that most public entities do not currently have available, ethically prohibit, or cannot deliver in an effective fashion—nor do most municipalities have a history of consistent social policy necessary to undertake such a problem. Since homelessness has been with us since the appearance of cities, perhaps we also need to consider its benefits.

THE SOUTH OF MARKET MILIEU

In 1980, the author moved into an industrial building with several artists relocating from the eviction of "squatting" artists of the Goodman Building located in the west Tenderloin/Central City district of San Francisco (Caldararo and Senger, 2012). This project followed an earlier one located in the Western edition. Artists were beginning to move to the South of Market area and out of the Haight, Castro, Fillmore, and Tenderloin districts due to increased pressure in housing costs spurred by renovations and real estate speculation. Between 1980 and 1989, a slow transition in occupation in the South of Market area was characterized in the press as its transition to a "live-work" ghetto for artists, but by the late 1980s, this was being transformed into a boom in renovations of the old warehouses into commercially available "live-work" condominiums occupied by professionals. Evictions of artists began to be a routine event as a truce between artists and the city's planners and public health authorities gave way to open condemnation of artist housing.

The 1989 earthquake put a temporary stop to this process only to resume with renewed vigor in the guise of determinations that many formerly occupied buildings by artists were now damaged and uninhabitable and unsafe. Many were tagged for demolition; others were abandoned. This began another transition where large numbers of transients and homeless individuals began to be present either by the result of the police enforcing vagrancy laws or due to the attractive nature of the abandoned buildings. Long-time artist residents and blue-collar workers responded to this change with some concern and mass meetings that were directed to fighting evictions and succeeded in designing temporary permits for artist live-work spaces. This gave way to addressing the issue of the new homeless.

As with the process of relocation of low-income residents that had been perfected by the Redevelopment Agency in San Francisco in other areas, the process in the South of Market area was organized and efficient with managed public relations, combined with community meetings by the Planning Department and the police to facilitate removal and relocation. Most of these meetings were advertised as informational to provide services and still held out the idea that some residents of buildings would be allowed to stay if certain code violations could be made.

In the midst of this transition, the earthquake provided an emergency that placed certain urgency on removal, this time under the guise that buildings were unsafe and, therefore, uninhabitable. Owners were given assurances that if buildings were damaged, permits and processing renovations and rebuilding of sites would be given priority. The removal was thus accelerated.

OBSERVATIONS OF HOMELESS AND BUSINESS/RESIDENT RESPONSE

Our observations from late 1980 to 1989 demonstrated a certain character to the homeless situation in South of Market. It was generally minimal in visibility, and there was a low impact on residents and business owners. The central research area was between Fourth Street and Fifth and Folsom and Harrison. The area was largely made up of two-story commercial buildings of brick or concrete with a few wood-frame buildings. There were no vacant lots with the exception of a parking lot on the corner of Fifth and Folsom on the north side that had been a parking lot during the entire study period (1980–2010).

Nearly half of the buildings housed sole owners' small businesses, though another 30 percent were partnerships with single-location businesses and the remainders were occupied by artists and a few outlets of larger corporations (e.g., a gas station). There were few vacancies, and artist occupation of buildings began to rise from 1970 to 1980 and reached a high point in 1985 when rents began to rise, especially long-terms rents signed in the late 1970s and early 1980s. The area was starting to attract businesses to its new "bohemian" atmosphere and cheap rent.

An unlicensed homeless center opened in the late 1980s in the middle of the block on the south side, but due to check-cashing fraud that brought nightly police intervention, it was quickly closed down. It temporarily increased the visibility of the homeless in the area, but this quickly subsided. From 1980 to 1989, the visibility of homeless individuals remained stable and local meetings of residents and businesses focused instead on problems surrounding housing and weekend violence and drunkenness of legal and illegal dance clubs and bars. Informal surveys (never taken with a clipboard or obvious writing instrument or notepad) demonstrated a number of "regulars" who moved from Market to Harrison where there was a freeway grassy area on a daily and weekly basis. Numbers ranged seasonally from under 10 in the winter to over 25 in the summer. Most of these people slept in the alleyways between Folsom, Harrison and Howard, either in nooks between buildings or in recessed fronts that provided some shelter from wind and rain.

Again, the homeless in the area had little visibility and interviews with business owners and workers reinforced residents' ideas that it was not a problem and that few homeless used the area. The fact that a regular group did use the area was not lost on many, but the idea of a general problem was absent. Some business owners provided water, sellable recyclable materials, an occasional meal, and some clothing to certain homeless individuals, while others simply ignored them. A few owners and residents repeated the idea that they had seen the homeless regularly pick up trash and dispose of it in proper receptacles. I observed this several times, but the same individual was responsible on each occasion. Another claim, less often reported in my surveys, was that homeless individuals were responsible for garbage: they spread it or set fire to it. I could not verify any of these claims but did find once a resident of the East Bay who stopped her car (a BMW) outside our warehouse at Fifth and Folsom and placed a large bag of her household garbage in our garbage can. She defended her action by saying that we had large garbage cans whereas she did not, and she had routinely done this for several years.

Another element of the neighborhood was the way economic behavior was organized between the homeless, the semi-homeless, and the residents. Some businesses functioned as banks, in that they cashed checks for people who had no bank accounts. Harvey's was run by a middle-aged Chinese-American businessman who had a long history in the city and whose children worked in his businesses. His main store was a liquor and grocery. He allowed the homeless and many residents and workers to establish "tabs" listing merchandise they had taken on credit. These were paid periodically. Harvey discounted interest on tabs depending on how people paid. Some made partial payments weekly, others monthly, and some asked for more credit to skip payments. People who did not have a bank account could also give Harvey money for safekeeping. He would offer chits or account letters of balances and deduct from them credit and purchases.

The bicycle messengers made up a large proportion of his clientele and many of these, as well as the homeless and other residents and workers, engaged in a vibrant local money-changing business where cash, checks, and food stamps could be exchanged. This was necessary as food stamps, for example, could not be used for some purchases, but by exchanging them for cash at a discount, money could be had to be used for prohibited objects. Since this took place outside of Harvey's, he neither encouraged nor discouraged it. When I once asked him about it, he simple shook his head and said, "People have to eat."

I was sure that Harvey took a number of losses with his customers and that the risk they represented also provided significant potential for gain in interest and charges or fees he collected. He often stored people's valuables for short or long periods for a fee and provided a message service, allowed use of phones, and other necessary and useful kinds of needed access. In all, Harvey was a bank, but his role was one that regular banks eschewed. By charging interest, he was able to cover his losses from bad loans, but extend credit and food to people who had no other recourse. He acted as both business and charity, and I often wondered how he balanced his books in the end. Many of the businesses in the area provided aid to the homeless, some at the initiative of employees, some by the owners or both.

AFTER THE EARTHQUAKE

On October 17, 1989, at 5:04 in the afternoon, the San Francisco Bay Area experienced a severe 6.9-magnitude earthquake. Buildings were severely damaged in several parts of the city, the Bay Bridge was damaged and closed, and some freeway sections collapsed or were damaged. Fires broke out in a number of places and falling brick superstructures killed some people. In our study area, the South of Market district, many buildings were so severely damaged that they were "red-tagged" (11) and could not be occupied; a few others were tagged as “yellow” (4), meaning occupation was limited until the buildings could be inspected. Another group was tagged “white,” meaning that they were judged to be undamaged and could be used.

Almost immediately, people tried to remove possessions from the buildings. In the first nights after the earthquake, we noted no other buildings with lights on the block. Our building was one of the six that were either residential or artist live-work spaces. None of the other residential buildings had occupants in the first week. In the second week, two other buildings were re-occupied by their tenants. A strange silence fell over the block day and night, broken only by trucks and cars arriving to remove possessions or attempt repairs. A considerable disagreement developed among the residents and the inspectors from city departments concerning the level of damage and safety.

In the second week, we began to notice a change in the homeless population. We had not seen any of our usual informants in the first week. In the second, some had returned but by the third week, we found that their numbers were increasing and the character of the homeless groups changed with larger numbers usually sporting their own “caravans” of 5 to 10 individuals of mixed sex and age lumbering by with shopping carts, dogs, and a few cats. Our usual population did not have more than one animal (a cat) in the nine years prior to the earthquake. Also, our usual population had only one child in this period, while the new groups had more. On one weekend in the March of 1990, I counted three different children in one afternoon.

The use of shopping carts seemed to appear as the local population of homeless was added to by those coming from across Market, that is, north to south. Interviews with a number of homeless who were new to us over the next 90 days indicated that police pressure was a factor in their movement but also that there seemed to be an increase in private security firms unfamiliar with the homeless, and these factors were also pressing the homeless to move from North Beach/Chinatown and the Financial District across Market.

Shopping carts seemed to be an adaptation to an expectation of movement for a longer time to greater, less secure locations. A number of these homeless individuals in groups moved through the South of Market farther south into the area along Fourth Street toward the Islais Creek area and beyond.

Prior to the earthquake, residents and business employees seldom reported car break-ins. After the earthquake, these increased dramatically. Many car owners sought garages to place their cars in both day and night. These were few and expensive. Residents began to complain to the homeless they knew about the situation, and a solution, mutually created, was slowly crafted. Some homeless people were found in cars in the morning, appearing to have broken into them. They argued that the cars had already been broken into and they just took the opportunity to get out of the weather.

On one occasion I stood by as a resident, furious that his car window had been broken, accused a homeless person of doing it in front of a policeman. The officer simply demonstrated how easy it was to break into a car and deftly picked up small rocks or pieces of metal to identify the size and shape and then tossed them against a wall to show how to throw the objects to produce the best results. I never saw anyone arrested for this crime. Eventually, residents came to accept the local homeless argument and simply left their cars open at night, allowing the homeless to sleep in their cars. The residents would then establish a time in the morning when they wanted to use their cars and would come down to find them empty. Some people rolled down all the windows in good weather, others put signs up stating, "Doors unlocked, do not break windows." Eventually, all the break-ins stopped.

An unofficial alliance of residents and workers developed where at night, homeless groups would arrange themselves in front of certain buildings acting as watch folk. Many had whistles around their necks, apparently taken up from the gay community's neighborhood watch program that gave out whistles to prevent attacks on people in the Castro at night. While I never personally heard one of these used at night, I was told of a number of instances when their use protected homeless people from abuse by drunks from local nightclubs and bars.

This unofficial watch extended in some areas, as long Fifth Street and Dore Alley to daytime. Many people reported, in one month, a number of instances of lost mail, and I myself saw our mail disappear after delivery when a man sitting in front of our door used a piece of metal with a wad of gum to fish out our mail from under the door and out the mail slot in the door. This rash of mail theft disappeared as soon as it appeared as whistles came to bear on attempts.

By late 1990, a number of empty lots had been created by the demolition of buildings. Some of these, as those at Fifth and Folsom, became the scenes of art events at night. Several groups of artists projected films on the walls of existing buildings at midnight or put on performance art pieces in them. Others set up sculptures, and all these lots saw homeless groups use them both during the day and at night. Artist use was usually limited to weekends and from 10:00 pm to 2:00 pm, leaving the sites available for homeless use.

Gardens also appeared. There was only one on the block at Fifth and Folsom. Homeless people, residents, and workers between Fifth and Fourth Streets organized several other gardens. These were short-lived in 1990 but reappeared in 1991, also briefly. Some people planted food; others unwanted houseplants or plants they found at other locations.

A more sinister result of the earthquake was the number of homeless and people using abandoned buildings for parties or illegal "raves." In 1990 and 1991, almost all the meetings with the police and the city staff were oriented to either close up the buildings or have them rehabbed or torn down as multiple fires spread across the Howard to Harrison areas with increasing frequency after the spring of 1990. Interestingly, only a few community meetings had been held in the area prior to the earthquake and these were organized by artists to try and legalize live-work space code changes.

By 1995, the complexion of the homeless situation had again changed, partly due to the change in occupancy of the buildings and the new construction. Rents were skyrocketing and artists were being moved out by eviction or rent increases. Small shops of blue-collar industries were also being priced out of the area. Interviews with new residents in the period 1992–1998 reflected a new population of young professionals and service industries, especially part of the dot.com boom. The arrangements with the homeless disappeared as the homeless melted away. While vacancies increased with the dot.com crash after 2000, there was no new visible population of homeless, though small numbers could be found on a daily basis. A community of residents, workers, and homeless had not survived.

Overall, the number of homeless seen on the streets has not changed and this appears to parallel the city in general. Problems that are often associated with homelessness, like panhandling, are still present in many parts of the city, but in our study area, the frequency of panhandling is about the same as it was in the 1980s. Another facet of homelessness as discussed above is the diversity of its performance. Many people cycle into housing and, even when in housing centers, spend their time on the streets in much the same way as when they are not residing in homeless shelters or city temporary housing. Kevin Fagan (2013) described this process in an article in the *San Francisco Chronicle*. As homeless numbers have dropped from a high of 8,640 in 2003 to a steady figure of about 6,500 over the past decade, the number of formerly homeless people in "supportive housing" have increased from 2,000 to 6,000, and the number of "supportive housing" units has increased from 1,000 in 2003 to 5,300 in 2013. This stasis is frustrating for city officials and homeless advocates, but may reflect a saturation point of homelessness in the San Francisco environmental context.

CONCLUSIONS

One might suggest that the common experience of the earthquake produced a form of communal spirit among the homeless, resident, and business employees. A kind of neglect produced by the economic effects of the earthquake left property available for a number of uses that were noneconomic, and the formation of bonds, temporary yet useful, resulted to produce reciprocal benefits. Economic pressures after 1992 began to eliminate part of these conditions and created new conditions that did not favor community formation.

This process of creation of community, whether South of Market in San Francisco in 1989 or 1906 or during encampments of Occupy Wall Street from Oakland to New York, represent some of the elements Tonnies (1887) described in the transition from stranger to acquaintanceship to community. In the case related in this article, the bonds of community between homeless people and residents and workers shifted from sympathy and antipathy to solidarity depending on the nature of issues impinging on the lives of individuals of each group. Parallels to this process are seen in the references above concerning established communities and the homeless in other locales and historic periods, as in the case of hobos.

The fact that no self-organized neighborhood groups appeared in the South of Market area we are concerned with in the entire period from 1980 to 2000 reflects the economic organization and lifestyle of the residents and workers. The day-to-night shift in population with most workers and owners living outside the area, while many artists and residents worked elsewhere, was not conducive to strong local community organizations. Previously, in the 1950s and 1960s, a powerful organization of residents, property and business owners and workers grew in the form of Tenants and Owners in Opposition to Redevelopment, but this organization became inactive by the 1980s (Hartmann and Averbach, 1974). Hartmann and Averbach attributed much of the success of Tenants and Owners in Opposition to Redevelopment to the large population of active retired former union members living in the area's single room occupancy buildings and union retirement housing.

Attempts by artists to organize code changes allowing live-work spaces to be legalized reflected this lack of organization. These efforts were compromised by builders and architects who used the issue to change codes to accommodate rehabilitation of older office buildings and warehouses into condos and rental lofts at prices most of the artists could not afford. This information is derived from interviews at meetings in the SOMA area in the period 1979 to 1990 and with San Francisco Planning Department employees.

Another example of the lack of organization and influence of the South of Market area was the fact that it was the only district of San Francisco that denied parking permits to residents. Meetings were organized by SOMA residents to establish such permits and permits were circulated, but no organization appeared to follow through the political process to achieve this end. My experience with other San Francisco neighborhoods demonstrated that success in this issue was largely dependent on the existence of community organizations with long-standing influence in local politics.

It is obvious that the reconstruction of modern cities around the world is resulting in systematic deviations from the space utilization described in

Narol's charts presented above. The use of open, public spaces is rapidly showing significant parallel form as shown in Harvey's (2011) samples, and the recent exposure of substantial private ownership of public spaces during the Occupy Wall Street and other locations outside of New York is another symptom.

Other locations show similar patterns, as from a short survey in Istanbul in 2010 showed a definite trend of demolition of smaller units occupied by working class Turks for larger expensive luxury units. Though in an interview by Dombey (2012) Abdullah Baysal, head of Istanbul's construction association reports on the increasing demand for large tower structures, while Vedat Asci, head of Astas Holding, the firm behind the new Macka Residences, argues that even when older housing is replaced, former tenants demand larger units than their old ones.

The only trend I have seen that does not conform to this increase in space is seen in preliminary surveys I conducted in Lisbon, where new constructions on the outskirts of the metropolitan area of Lisbon have provided units affordable and consistent with small family units typical of earlier construction. India does not fare much better, where estimates of cities like Mumbai find that at least half the population lives in slums of makeshift structures, most unlikely to survive severe weather (Crabtree, 2012). In the case of these structures, whether in Mumbai or a favela in Brazil, the difference between "homelessness" and living in an illegal structure on private or public property is a narrow slice of reality, not just due to eviction when development takes place, but given the septic conditions and poor protection from the elements.

Recent works by Edward Glaeser (2011) and Stewart Brand (2010) have celebrated slum housing, ignoring the suffering of the population, the poverty traps and lack of services including education, and have instead focused on the "entrepreneurial" opportunities of "self-made" living conditions. Kotkin (2005) has argued that the failure of cities in the past to provide security for all their residents has been a significant factor to their disintegration and collapse. In these definitions of city life, the idea of what is a home and what constitutes a living space come into focus as blurred visions of simple survival for the majority of the world's population. David d'Heilly (Pilling, 2011) argues that cities are no longer recognizable, but rising and expanding wherever people can survive long enough to produce and consume. This future looks much more like the medieval city described by Sjoberg (1960) than the ordered cities of Sumer and Mari that have defined our ideas of city life for over 4,000 years (Lambery-Karlovsky, 1974).

But what is of interest regarding our study and these theoretical visions of the city is the fact that out of the uncertainty of the earthquake, homeless and residents created forms of community in common spaces. The autonomous nature of these associations formed services of a reciprocal exchange. These associations and services continued as long as the spaces were unclaimed and resident stability provided continuity. Recent research I have conducted in San Francisco indicates that where local residents have been given control of neighborhood parks (specifically Brooks Park), they have organized and built gardens, controlled graffiti, and policed litter. This is the same conclusion Schuessler (1992) found in her study of parks in Providence.

So, humans in this chapter, and in the study outlined, can be "homeless" to varying degrees, yet form cooperative relations based on their different situations. The pressure to "house" the homeless continues, however, with those without stable locations seen as problems. This pressure seems to be an artifact of the ideals of modernity projected onto places claimed by the "homeless" as their temporary homes: parks, sidewalks, abandoned buildings, and so on. Those who are homeless are also economic members of lower-earning classes or individuals made abnormal by different processes of human activity and biology: war veterans (they make up a large number), drug and alcohol "abusers" (really those whose biology or life trajectory have left them victims of social competition). Perhaps our main problem is too many people, and not too little space. A reflection on an earth with a human population of one billion might be more sustainable in its demands on resources and, given our present technological expertise, provide greater rewards for those humans and nonhumans sharing the planet.

NOTES

1. See "Five percent of the nation's housing units qualify as second homes. But where the hec are they?" SecondShelters.com, accessed March 14, 2017, http://secondshelters.com/2011/08/29/percent-nations-housing-units-qualify-homes-hec-they/
2. See "Historical census of housing tables," U.S. Census Bureau, accessed March 15, 2017, http://www.census.gov/hhes/www/housing/census/historic/vacation.html

Chapter Thirteen

On the Curious Illusion of Human Uniqueness

> The subtle way of the universe appears to lack strength, yet its power is inexhaustible.
>
> —Lao Tzu, sixth century BCE

It is strange that humans trace their beginnings to the asteroid that demolished the world of the dinosaur. Our ancestor in that pre-apocalyptic world was a tiny insectivore, a nocturnal mammal whose perceptions were formed more by smell than by the vision that dominates our cognition today. But in searching for our origins, the path often comes to define the seeker more than the home or the future. Or, in our guise as a stranger in this new world and a refugee of the old, it is as if "when (one) makes his appearance among people who do not know him, the only pedigree (nasab in the sense of nisba) he can muster is one that connects him to the road whence he comes" (Rosenthal, 1997).

One might suggest that the common experience in the pre-earthquake period produced a form of communal spirit among the homeless, residents, and business employees—a kind of benign neglect resulting from the economic conditions. The effects of the earthquake created a reduction of this benign neglect and community context. The "road" of evolution seems convoluted and our pedigree muddied. The future and the demands of technology loom over us, creating landscapes where no animal has walked or will walk. While we began this book with an appraisal of the evolution of the brain, we have traced the benefits of the behavioral consequences over time and through the appearance of complex behavior. One aspect of this complexity has been a number of forms that some humans have generally found to be unsavory, like war. War appears in other animals with complex societies, as in ants. But what is remarkable in our survey has been the continuous nature of irrationality or illogic in human behavior and institutions. Some might argue that irrational behavior can be adaptive and produce outcomes that are beneficial to individuals and their societies. Do we dare consider that our walk into the future is driven by illusions and irrational views of ourselves?

E.O. Wilson has argued (1975) that humans are easily "brainwashed" and mass action by members of human societies also produce distasteful consequences, as in mob action, the Inquisition, and racist pogroms. Canetti (1966) found that the formation of mobs was not rare and yet the participation of individuals in them quite common. Membership in groups as in mobs gives protection and is contrasted with the outsider, the stranger, the outcaste. For Mary Douglas (1966), symbols of pollution and decay can be associated with survival and mark the Other, as in Freud's concept of the threat and motivation of fear of death. Yet, these concepts are tinged and structured by culture and vary over time and peoples.

One of the most profound contradictions in human behavior is how the mass of humanity can act on ideas that are fantastic, absurd, or patently false. Often, highly educated people can be motivated to follow such ideas as in Nazi Germany, the racist wars of the former Yugoslavia, and Rwanda. Freud (1966) was struck by the history of religion's power and realized that it could be a powerful force for unity and cooperation, but also for terrible punishments, mass murder, and horrific acts of torture. But it is often joined with other concepts that are equally potent motivators, such as racism, nationalism, and ideologies like capitalism and communism.

BOMBS, DÉTENTE, AND HUMANS LIKE ANTS

It is interesting, in the present environment of attitudes of human exceptionalism and uniqueness, that millions of years ago some social animals reached the level of development of chemical warfare that humans have achieved only in the past 200 to 100 years (Wilson, 1975). Ants and bees engage in cleaning of their colonies, and ants inoculate each other when disease is present and practice avoidance of ill colony members—behaviors lacking in some human societies (Caldararo, 2012b, 2015a).

Once sedentary behavior became a median mode of living for humans, we find human structures paralleling those in which domesticated animals live. Today, huge factory farms look like high-rise buildings or human prisons. We now live like our food does with regular feeding schedules and exercise. In some of the specialized soldier castes of ants and termites, not only have soldiers become adapted in the variation of their physiology to block entrances to nests as in *Pheidole lamia* and *Cephalotini* species, but also in some the head shape has become specialized to cutting or crushing the enemy as in *Camponotus*. Others have sickle-shaped or hook-shaped mandibles to pierce the bodies of foes as in army ants and driver ants.

Most impressive is the adaptation of the soldiers of the Australian termite *Mastotermes darwiniensis*, whose glands produce p-benzoquinone in the mouth cavity and, mixed with saliva, amino acids create a rubberlike material that entangles the enemy troops. In soldiers of the *Globitermes sulfureus* we have walking "time bombs," as E.O. Wilson (1975) describes them with a conical organ at the nose area. Other species of termites have evolved in similar means by convergent evolution as in the *Nasutitermitinae*, who can shoot their chemical bomb by powerful mandibular muscles over several centimeters, and its aim is amazing given they are blind.

The fact that these defensive and offensive systems evolved over millions of years to reach a rather interesting level of "détente" gives rise to speculation on human systems of aggression. Will we continue to develop more lethal ones that will threaten all life or will we reach some point of balance like ants and termites? While we do not possess the same learning and intrinsic systems of information and response as these social animals, one might imagine that such "détente" is a sign of successful mature social evolution in animals. In his book *Nuclear Weapons and Foreign Policy*, Henry Kissinger writes, "We fear force as never before, and we even fear economic and political measures which might lead to the use of it." Today, with the rise of China and the decline of American industry (largely through off-shoring and over-seas investment as well as a program of transferring government assets to entities abroad not subject to U.S. government control or taxation as for example, the sale of the U.S. merchant marine fleet after World War II), the economic system is unstable, religious fanaticism and nationalism are returned threats. Humans face the challenge of surviving the products of our defensive abilities and our fears of each other. This is probably the greatest challenge of all.

The current currency crisis is built upon this process of divestment of U.S. assets abroad and the transfer of trillions of dollars in wealth beyond U.S. tax authorities. The media talks of a "currency war" and argues that China is the aggressor and manipulating its currency. This is rather silly, since China has simply spent the past 20 years being better capitalists than the West. What is being built is a blaming of the scapegoat for the results of the 2 percent of American wealthy, who have profited from the destructive tax policies and wealth transfer process (off-shoring, etc.) of the past 50 years. So, does war substitute for business or vice versa? Or are these actions part of a general behavior of spite as Wilson (1975) has suggested in several contexts for animals? Does war and business exercise or satisfy needs and also that of spite? This is another problem with regarding human behavior as rational.

The contradiction of education and rationalism being subject to the whims of irrationality forces us to consider that the human brain is like a computer, "garbage in, garbage out," as the saying went in computer science. Yet, it is dissatisfying to conclude on such a point. We often retreat from such stark evidence of a lack of morality and ethics by emphasizing unique behaviors like love, altruism, music, and the arts in general. We attempt to clothe human failures in the ideals and not the real. But this brings us to the issue of knowledge and generations in a different sense.

KNOWLEDGE AND POWER

Gillian Tett's article in the *Financial Times* of February 15, 2013 ("Library books are on borrowed time"), reviews some of the challenges facing libraries today. We should recall that libraries have been of great utility in all the major civilizations since the Egyptians, Sumerians, and Chinese emerged from prehistory. While debates swirl about the nature of literacy in the Indus and Peruvian civilizations, it is clear that books were central to learning among the Mayans and Aztec. Ernst Posner's *Archives of the Ancient World* and G.R. Driver's *Semitic Writing* trace the advent of reading and libraries from Sumer to Greece and Rome. We know that the libraries of Carthage contained the records of the most ancient voyages of the Phoenicians, whose details are now lost, though the Romans gave the books to their African allies, King Jurgutha and King Juba. The loss of books and knowledge has been constant throughout history (Caldararo, 1994).

Julius Caesar established the first free public libraries in Rome, and later emperors and benefactors added to these. Libraries functioned in the Roman Empire as centers of business and learning for another 500 years until the collapse of Roman order. C.E. Boyd (*Public Libraries and Literary Culture in Ancient Rome*, 1915) describes how the public used the libraries of the classical Roman period and how they were designed. A thousand years passed before public libraries again appeared in Europe in the Renaissance.

While Dr. Tett cites authorities that argue books and libraries are passé, we see them appear and disappear in human history as change challenges societies in various forms. It is unlikely, however, that many books produced on electronic platforms would survive a social cataclysm similar to the fall of Rome. Books survive chaos due to their popularity and numbers, or the materials they are made of (e.g., cuneiform tablets), and often due to the inaccessible ways they were hidden or stored (Nag Hammadi library, Cairo Geniza). One can hardly imagine someone being able to "read" a kindle in the twenty-third century after a social collapse; it would be like today trying to read a pre-Columbian quipu or an IBM punch card or floppy disk. While technology changes, our desire to transfer knowledge to future generations who will be in need of past wisdom will not change. Philip Pettifor (2005) (of Libri) has argued that libraries have been forced over the past two decades to become social welfare institutions and not just places where books can be read or lent. Instead, they are refuges for the homeless, the unemployed, the disabled, and children whose parents both work; for these patrons, they are forced to perform a variety of other duties, compensating for declining social services and the budget choices of cash-strapped working- and middle-class families across Europe and America.

Also, sadly, the hubris of every generation to reject the past for the new seems also to be unchanging. Libraries today are suffering more due to budget cuts than a lack of interest, and closing them is short-sighted and detrimental to the development of habits of youth for knowledge. The transition from generational face-to-face teaching combined with the abandonment of the book is most likely the greatest change in human interrelations since the appearance of literacy (Goody, 1987) and is likely to have as dramatic social consequences as did contact in Melanesia as Mead (1956) describes it.

Fortunately or unfortunately, humans are not born with knowledge; we must learn and forget. We have decided to trust in technology to provide for a future's need for knowledge. How would humanity deal with a future in which all knowledge disappeared with the exception of the memories of adults? Could science and technology be recreated? Without manuals for making cars, engines, chemicals, drugs, for teaching medicine, and so on, could humanity reinvent what was needed to survive an electronic catastrophe? Perhaps a Zika-like virus infection would create humanity without knowledge for decades. Could we recover? Could it be done in time to avert disasters and famine, and to fight disease? Can we plan for that?

DISASTER AND THE ANTHROPOCENE

While we fear economic collapse and watch movies that investigate social collapse, we seldom consider the difficulties of rebuilding or renewal after a disaster more brutal than the 1929 economic depression or the Black Plague. Studies of renewal have been undertaken in the past, yet most have been limited in area or regional in focus. A worldwide renewal study is rare and most attempts have been more fantastic than sober. We have been treated to a plethora of scenarios of climate disasters in recent years, and new definitions of the effects of the human environment (the Anthropocene) are chilling. Chew (2007), for example, argues that collapse and "Dark Ages" have a renewal function, given the degradation of the environment resulting from human accumulation of wealth in all its forms, both physical and symbolic. I have shown elsewhere (Caldararo, 2013), however, that sustainability and stability are possible without collapse dependent on cultural proscriptions against certain levels of accumulation that would lead to degradation and collapse. The ancient Khmer appear to be an example of this variation (Caldararo, 2015).

The missing element to the other theories of collapse is their failure to account for this variation, especially the role of ideologies. Rostovtzeff (1926) suggests that Neo-Pythagoreanism was at the heart of the Roman economic expansion after the civil wars and was a central element in rebuilding the stability of the Empire. Reischauer and Fairbank (1960) pursue a similar role for Neo-Confucianism in the dynastic form in China and I have argued that a culture that has integrated Shinto, Neo-Confucianism, and Buddhism has allowed Japan to produce a resilient response to challenge as well (Caldararo, 2003b, 2016a). The Japanese have responded to environmental stress that reduced resource availability (e.g., food) several times by the reduction of population. The country is undergoing a current period of population decline, which could reduce its resource extraction to a closer frame of carrying capacity as in the past only with now industrial technology. The rest of the world's current economic system is based on an ideology of capitalism requiring constant growth and consumption and its promise of riches, but it functions on magical thinking and irrationality that also promotes inequality and environmental degradation as the author has shown in a number of publications recently (Caldararo, 2004a, 2009b, 2009c, 2011). We are experiencing a rapid change in the environment; die-offs of amphibians and pollinators (Chapin et al., 2000) as well as the pollution of the seas and depletion of fish (Palumbi, 2001) seem to bring us to the conclusion that complex society, in vertebrate form as produced by *Homo sapiens*, is not sustainable, or that its variable survival might pattern in similar fashion to the lifetime of a specific bee hive or ant colony and then require renewal in some form or, worse, extinction (Boulter, 2002).

Regeneration after collapse is often affected by climate, especially in some cases as in Mayan patterns (Marcus, 1998). Regeneration takes a variety of forms dependent on local environment, extent of collapse, intrusive peoples, and type of complexity of the original society and its integration (Blanton et al., 1996). Stark (2006) has emphasized the idea of regeneration and movement in the Khmer example with the idea that the continuity of symbols and institutions and an Indic ideology allowed for success. The failure of regeneration in the case of the influx of a powerful new culture and society coupled with efforts to prevent continuity of native symbols and institutions can be seen in the case of Native American complex societies where destruction of Native architecture, writing, learning, and institutions of nearly every type created a permanent hiatus concerning Native complex culture (Caldararo, 1994).

While it is obvious from the fossil record that climate change has occurred in the past, the factors involved—reduced solar radiation, Milankovitch cycles of axis variation in tilt, albedo effect, and the influence of life processes—are many. It is clear, nevertheless, that life processes have an effect. The Great Oxidation Event that occurred about 2.3 billion years ago was the result of life processes. Cyanobacteria had been pumping out oxygen for more than a billion years, and apparently, this accelerated between 2.4 and 2.3 billion years ago dramatically changing the atmosphere of the earth (Margulis, 1981, 1992) and clearly toxic conditions to anaerobic forms.

Loss of diversity at the end of the Devonian has been explained by a reduction in speciation and invasions of cosmopolitan species, rather than any single event, but each extinction period is associated with anoxic sea conditions that may be the result of volcanic action or biological activity by certain species and their metabolites. While other factors may play more important roles in certain extinctions and changes in climate, for example, possible asteroids in the Permian extinction or at the end of the Cretaceous, it is clear that life processes do have a role in climate change.

CLIMATE CHANGE AND POPULATION PRESSURE

The central questions then are the children and the aged, as we have to stabilize our population and the impact we have on the environment. In Japan, the problem, as I have noted, is already in solution. The demographic transition is underway for its aging population, as Japan's younger generations are actively caring for their parents, and there is a lively health care industry catering to the aged. Ellen Freudenheim has written on how advanced the Japanese health care industry is regarding the elderly and robotics (http://suite101.com/article/service-robots-as-caregivers-for-frail-elderly-a153372).

Human population does not just need to stabilize at 8, 10, or 12 billion, but it also needs to be reduced to perhaps 5 or 6 billion by 2050, and 2 to 3 by 2100. We must ask ourselves what good is a continuing population increase. This debate has been continuous over the past 2000 years and likely before (Hutchinson, 1967). With a decreasing pressure from population and pollution, along with some innovations for robotics, a future for humanity is possible. The question of children is intertwined here. It is not how many children as the question was for China, but what kind of life children have.

In 2014, in an article by Janan Ganesh ("Being childless," *Financial Times*, June 28, 2014), he approaches the issue from the perspective of the simple production of children and how society regards childless people. On the other hand, if we consider the value and nature of children to society, we find a different set of problems. The work of two French scientists, Philippe

Aries (*Centuries of Childhood*, 1962) and Lloyd deMause (editor of a volume entitled *The History of Childhood*, 1974), argues that we do not need children today, that they have no real role in life. In the past, children had economic roles; they gathered food, watched animals, worked in factories. Today, we exist in a period, as Hugh Cunningham, author of *The Invention of Childhood* (2006), writes, where there is no social product for children to produce, and so, we have created "activities" that are meant to surround the general acquisition of skills defined in education as meaningful. If we look to the nature of variations in the quantity of children in different societies over time and their value, we find significant outcomes. The example of Florence in the Renaissance is a good one. Using existing data, David Herlihy in 1977 showed how rates of marriage changed between 1300 and 1500. If we look at the year 1427, we find that the average age of first marriage in the Florentine countryside is much lower (25.6 years) to that in Florence (30.0 years). The percentage of married (adults) in the countryside was 66.8 compared to only 47.6 in Florence, and the percentage of permanent bachelors was 5.4 in the countryside and 11.8 in Florence. When we look at families with male-only or female-only heads, we find another correlation with urbanization. In 1427, male-only headed households were 26.3 percent of the population while female-only ones were 15.4 percent, but in the countryside, male-only households were only 13.4 percent and female-only at 9.1 percent.

The city and country female-only families may reflect what sixteenth-century writer Lodovico Dolce said about widows, that some women rejoiced at the death of their husbands, "As if a heavy yoke of servitude had been lifted from their backs." An additional finding by Herlihy, which provides us with an insight into the unique economic and social organization of Florence at this time, was that the highest fertility rate was among both the richest urban and rural Florentines. The middle class and poor had fewer children. I am not advocating such a trend, but it is clear that fewer, better educated, and better fed children are perhaps a better starting point for dealing with present population and resources. It might result in a twenty-first-century Renaissance.

The issue is really not how many children, but the kind of children we see for the future, ones in a cheek-by-jowl world, or children who will be raised in a time of transition for the next Renaissance? Social engineering is always a curious and suspect venture, but we have engineered our cities, roads, food, banks, and money, and yet, planning for an earth with fewer people might be a subject to consider. Yet, children are economic drivers of consumption, but their consumption, while affected by media in advertising, is still largely mediated by their parents' income and preferences.

It is always confounding when people assert that we are better off today than in the past. When one seriously addresses this problem, as I did in my 2013 book and a recent article (Caldararo, 2016c), it is not as clear as is often asserted. We cannot project ourselves back 10,000 years to a time when all humanity was still hunting and gathering. While archaeology has offered impressions of that life, ethnographic views of more recent examples surely cannot be used as authentic comparisons, reduced as they became. In the same fashion, claims that billions have been raised out of poverty in the past 40 years are absurd (Caldararo, 2016c). What has happened is that hundreds of millions have been forced into cash economies, so they went from being self sufficient to earning $2 a day.

The question for people today is one of consciousness and what it means to be human. We often pride ourselves by believing we are thinking animals, so one might expect that, given past roles, "dumb" animals and plants have played in climate change, should we not recognize our own role and behave accordingly? Will we create conditions that are toxic to our own survival, as other life forms have done in the past?

Steven Hawking believes the universe is rational (Hawking and Mlodinow, 2010), but since humans are seldom rational and often emotional, can we ever "understand" the universe? Etzioni (1988) argues that group decisions often produce more balanced and useful decisions in some cases, one might say, causing an integration of rationality, consideration, and emotion. While the idea of a rational universe is somewhat comforting, it is reminiscent of religions that posit moral gods, as in Christianity, who are more like tribal authorities that execute justice and morality at whim (or as with Job in the Bible, act illogically) as opposed to the polytheistic gods of Greece and Rome or Hinduism, who have no pretense to morality or justice and act like humans, raging, acting in indulgent ways in pleasure or war. The answer to the question of why gods act immorally or unjustly is that their ends are mysterious to us—the same kind of answer given to children about adult misbehavior or contradiction.

All this depends on the human brain in the end, and yet our brain may be the enemy of a successful and peaceful future. While humans created culture, Leslie White (1949) recognized that culture makes humans and outlives each person who possesses it; however, its existence and function may no longer promote the lives of humans, but of the society it also created, as Jules Henry feared in his book *Culture against Man*. On the other hand, considering the greater social mind of culture and our genetic load (lack of fitness for survival, defined as detrimental genes, mutations, or traits, see Moody, 1967), the combination has created the superorganism.

The human superorganism is in its initial stages of development, and it is unlikely to look as the most developed economies today, or any eusocial invertebrates, do. Should population stabilization take place in the next 100 or 1,000 years, as some demographers predict, forms of control of reproduction will have significant repercussions on social life, as will means of social control to adapt mammalian life to much greater density and loss of space than is common today. On the other hand, if either conscious or accidental (pathogenic) population reduction occurs and human density falls to pre-industrial levels, the benefits of technology might create a variety of potential human social organization, perhaps even more varied than has occurred thus far, and perhaps not superorganistic. These scenarios depend on the organization and development of the superorganism; given that invertebrate eusociality is millions of years old, should humans survive, superorganism evolution might require tens of thousands, if not hundred thousands of thousands, of years to create a stable system.

The goals and future of that "organism" are not different from those of colonies of yeast or ants, and no matter what great strides we have made in technology, from stone tools to biotech, even if we achieve cloning humans or colonizing new planets, we are doing nothing different in kind than what yeast does: reproduce and colonize.

Source: © Dan Piraro, 2016. All rights reserved. Reproduced with permission.

References

Abi l-Hadid, Ibn, *Sarb Nahg al-balaga,* edited by Hasan Tarim (Beirut, 1963–1964).

Abraham, Carolyn, "Are we treating a disorder or boyhood as a disease?" *The Globe and Mail*, October 18, 2010.

Adams, Robert McC., *The Evolution of Urban Society* (Chicago: Aldine, 1966).

Adams, Robert McC., *Heartland of Cities* (Chicago: Chicago University Press, 1981).

Adisetiyo, Vitria, Jens H. Jensen, Ali Tavesh et al., "Multimodal MR imaging of brain iron in attention deficit hyperactivity disorder: a non-invasive biomarker that responds to psychostimulation treatment?" *Radiology*, 272, No. 2 (2014): 524–532.

Aiello, Leslie C., "Brains and guts in human evolution: The expensive tissue hypothesis," *Brazilian Journal of Genetics* 20, No. 1 (March 1997). doi: http://dx.doi.org/10.1590/S0100-84551997000100023.

Aiello, Leslie, and Peter Wheeler, "The expensive tissue hypothesis," *Current Anthropology* 36 (1995): 199–221.

Alexander, Richard D., "Evolution of the human psyche," in *The Human Revolution*, edited by Paul Mellars and Chris Stringer (Princeton: Princeton University Press, 1989): 455–513.

Allday, Erin, "A few more homeless, not as many on streets," *San Francisco Chronicle*, April 5, 2009.

Allee, W.C., *The Social Life of Animals* (Boston: Beacon Press, 1938, 1958 edition).

Allee, W.C., *Cooperation Among Animals* (New York: Schuman, 1951).

Almond, Gabriel, M. Chidorow, and R.H. Pearce, editors, *Progress and Its Discontents* (Berkeley: University of California Press, 1982).

Anderson, Britt, "The g-factor in non-human animals," *Neurology* 127 (January 2000), Novartis Foundation symposium 233: 79–90; discussion: 90–95.

Anderson, B., "Naltrexone (NTX) ameliorates the behavioral consequences of methylazoxy-methanol (MAM)-induced microcephaly," *Neurology* 44 (1994): A412.

Anderson, Denis, and Iain J. East, "The latest buzz about Colony Collapse Disorder," *Science* 319 (February 8, 2008): 724–725.

Anderson, Nels, *The Hobo: The Sociology of the Homeless Man* (Chicago: University of Chicago Press, 1961, first published in 1913).
Anderson, Robert, *The Aims and Achievements of Medical Anthropology: Magic, Science and Health* (Fort Worth: Harcourt Brace, 1996).
Anderson, R.M., and R.M. May, "Population biology of infectious diseases: Part I," *Nature* 280 (1979): 361–367.
Anderson, Roland, Jennifer Mather, and James B. Wood, *The Ocean's Intelligent Invertebrate* (Portland: Timber Press, 2010).
Andersen, Sandra B., Sylvia Gerritsma, Kalsum M. Yusah et al., "Natural history note: the life of a dead ant: the expression of an adaptive extended phenotype," *American Naturalist* 174, No. 3 (September 2009): 424–433.
Anderson, J., S. Williams, R. McGee et al., "DSM III disorders in preadolescent children," *Archives in General Psychiatry* 44 (1987): 69–76.
Andics, A., A. Gabor, M. Gacsi et al., "Neural mechanisms for lexical processing in dogs," *Science* 353, No. 6503 (September 2, 2016): 1030–1032.
Andics, Attila, Márta Gácsi, Tamás Faragó, Anna Kis, and Ádám Miklósi, "Voice-sensitive regions in the dog and human brain are revealed by comparative fMRI," *Current Biology* 24, *No. 5* (February 20, 2014): 574–578.
Andrews, Edward, "Christian missions and colonial empires revisited: a black evangelist in West Africa, 1761–1816," *Journal of Church and State* 51, No. 4 (2010): 663–691.
Annis, Barbara, and John Gray, *Work with Me: Gender Intelligence Can Help You Succeed at Work and in Life* (London: Piatkus St. Martin's Press, 2013).
Ansorge, M.S., Mingming Zhou, Alena Lira et al., "Early-life blockade of the 5-HT transporter alters emotional behavior in adult mice," *Science* 306 (2004): 879–881.
Anthony, David W., *The Horse, the Wheel and Language* (Princeton: Princeton University Press, 2010).
Appian, 165 A.D. *The Civil Wars*, translated by John Carter (Harmondsworth: Penguin, 1996).
Araque, A., V. Purpura, R.P. Sanzgiri, and P.G. Haydon, "Tripartite synapses: glia, the unacknowledged partner," *Trends in Neuroscience* 22 (1999): 208–215.
Arbib, Michael A., *How the Brain Got Language* (Oxford: Oxford University Press, 2012).
Arch, Victoria S., T. Ulmar Grafe, Marcos Gridi-Papp, and Peter M. Narins, "Pure ultrasonic communication in an endemic Bornean frog," *PLoS ONE* (April 29, 2009). doi: 10.1371/journal.pone.0005413.
Arendt, Detlev, Maria Antonietta Tosches, and Heather Marlow, "From nerve net to nerve ring, nerve cord and brain – evolution of the nervous system," *Nature Reviews: Neuroscience* 17 (January 2016): 61–72.
Argue, D., D. Donlon, C. Groves, and R. Wright, "*Homo floresiensis*: microcephalic, pygmoid, *Australopithecus*, or *Homo*?" *Journal of Human Evolution* 51, No. 4 (October 2006): 360–374.
Arnold, Thomas Walker, *The Preaching of Islam: A History of the Propagation of the Muslim Faith* (London: Constable, 1913).
Asander, Harje, *A Field Investigation of Homeless Men in Stockholm: A Socio-Psychiatric and Clinical Follow-up Study* (Copenhagen: Munksgaard, 1980).

Aspden, Peter, "The rules of abstraction," *Financial Times*, Weekend Edition, March 18, 2006.
Aviles, Leticia, and Wayne Maddison, "When is the sex ratio biased in social spiders? Chromosome studies of embryos and male meiosis in *Anelosimus* species (Araneae, Theridiidae)," *Journal of Arachnology* 19 (1991): 126–135.
Azevedo, Frederico, Ludmila R.B. Carvalho, Lea T. Grinberg et al., "Equal numbers of neuronal and nonneuronal cells make the human brain an isometrically scaled-up primate brain," *Journal of Comparative Neurology* 513 (2009): 532–541.
Bachofen, J.J., *Myth, Religion and Mother Right, Selected Writings,* translated by R. Manheim (London: Routledge and Kagan Paul, 1967). Originally published as Das Mutterrech, 1861.
Backus, Emily, "Armchair psychology," *Financial Times* (2009), House and Home Section: 1.
Baddeley, A., "Working memory," *Science* 255 (1992): 556–559.
Balter, Michael, "Cave structure boosts Neandertal image," *Science* 271, No. 5248 (January 26, 1996): 449. doi: 10.1126/science.271.5248.
Balthazart, Jacques, Thierry D. Charlier, Jennifer M. Barker et al., "Sex steroid-induced neuroplasticity and behavioral activation in birds," *European Journal of Neuroscience* 32, No. 12 (2010): 2116–2132.
Bamshad, Michael, Toomas Kisisild, W. Scott Walkins et al., "Genetic evidence on the origins of Indian caste populations," *Genome Research* 11, No. 6 (June 2001): 994–1004.
Baptist, Edward, *The Half Has Never Been Told* (Boston: Basic Books, 2012).
Barak, Gregg, *Gimme Shelter: A Social History of Homelessness in Contemporary America* (New York: Praeger, 1991).
Barber, E.J.W., *Prehistoric Textile, Development of Cloth in the Neolithic and the Bronze Age* (Princeton: Princeton University Press, 1991).
Barnard, Alan, "When individuals do not stop at the skin," in *Social Brain: Distributed Mind,* edited by Robin Dunbar, Clive Gamble, and John Gowlett (Oxford: Oxford University Press for the British Academy, 2010): 249–267.
Baron, Georg, Heinz Stephan, and Heiko D. Frahm, *Comparative Neurobiology in Chiroptera, V. 1 and 2* (Basel: Birkhauser Verlag AG, 1996).
Barro, Robert, Jason Hwang, and Rachel McCleary, "Religious conversion in 40 countries," *Journal for the Scientific Study of Religion* 49, No. 1 (2010): 15–36.
Barron, Andrew B., Ryszard Maleszka, Paul G. Helliwell, and Gene E. Robinson, "Effects of cocaine on honey bee dance behavior," *Journal of Experimental Biology* 212 (2008): 163–168.
Barton, R.A., "Neocortex size and behavioral ecology in primates," *Proceedings of Royal Society of London B* 263 (1996): 173–177.
Barton, R.A., "Embodied cognitive evolution and the cerebellum," *Philosophical Transactions of the Royal Society B* 367 (2012): 2097–2107.
Barton, Michael C., Julien Riel-Salvatore, John M. Anderies, and Gabriel Popescu, "Modeling human ecodynamics and biocultural interactions in the Late Pleistocene of Western Eurasia," *Human Ecology* 39, No. 6 (December 2011): 705–725.
Bar-Yosef, O., and A.J. Belfer-Cohen, "The origins of sedentism and farming communities in the Levant," *Journal of World Prehistory* 3 (1989): 447. doi: 10.1007/BF00975111.

Barzun, Jacques, *Race: A Study in Modern Superstition* (New York: Harcourt, Brace, 1937).

Basu, Analabha, Namita Mukherjee, Sangita Roy et al., "Ethnic India: a genomic view, with special reference to peopling and structure," *Genome Research* 13 (2003): 2277–2290.

Bateson, Gregory, *Steps to an Ecology of Mind* (Berkeley: Ballantine Books, 1972).

Batra, Suzanne W.T., "Behavior of some social and solitary Halictine bees within their nests: a comparative study (Hymenoptera: Halictidae)," *Journal of the Kansas Entomological Society* 41, No. 1 (January 1968): 120–133.

Baxter, Peter Stuart, Alan Steven Rigby, Marianne H.E.P.D. Rotsaert, and Ingram Wright, "Acquired microcephaly: causes, patterns, motor and IQ effects, and associated growth changes," *Pediatrics* 124, No. 2 (August 1, 2009): 590–595.

Beari, Maddalena, and Craig Stanford, "A bigger, better brain: observations of chimpanzees and dolphins strengthen the notion that humanlike intelligence may not be uniquely human," *American Scientist* 98, No. 5 (September–October 2016): 402–409.

de Beauvoir, Simone, *The Second Sex* (New York: Vintage Books, 1952).

Bednarik, Robert G., "A figurine from the African Acheulian," *Current Anthropology* 44, No. 3 (2003): 405–413.

Beier, A.L., *Masterless Men: The Vagrancy Problem in England 1560–1640* (London: Methuen, 1985).

Bekoff, Marc, "Animal emotions: Do animals think and feel?" *Psychology Today*, August 12, 2013.

Bellah, R.N., "Religious evolution," in *Reader in Comparative Religion: An Anthropological Approach*, 4th edition, edited by W.A. Lessa and E.Z Vogt (New York: Harper and Row, 1978): 36–50.

Benavides-Piccione, R., I. Ballesteros-Yanez, J. DeFelipe, and R. Yuste, "Cortical area and species differences in dendritic spine morphology," *Journal of Neurocytology* 31 (2002): 337–346.

Benyamin, B., B.S. Pourcain, O.S. Davis et al., "Childhood intelligence is heritable, highly polygenic and associated with FNBP1L," *Molecular Psychiatry* 19, No. 2 (February 2014): 253–258.

Berger, Lee R., John Hawks, Daryl J. de Ruiter et al., "*Homo naledi*, a new species of the genus *Homo* from the Dinaledi Chamber, South Africa," *eLife* 4 (2015): e09560.

Bergman, O., L. Westberg, L.G. Nilsson, R. Adolfsson, and E. Eriksson, "Preliminary evidence that polymorphisms in dopamine-related transcription factors LMX1A, LMX1B and PITX3 are associated with schizophrenia," *Progress in Neuropsychopharmacology & Biological Psychiatry* (June 4, 2010). Epub ahead of print.

Berman, Bruce, "Ethnography as politics, politics as ethnography: Kenyatta, Malinowski, and the making of Facing Mount Kenya," *CJAS* 30, No. 3 (1996): 313–344.

Berna, P., P. Goldberg, J.S. Brink et al., "Microstratigraphic evidence of in situ fire in the Acheulean strata of Wonderwerk Cave, Northern Cape province, South Africa," *Proceedings of the National Academy of Sciences, USA,* 109 (2012): E1215–1220.

Berreman, Gerald D., "Scale and social relations," *Current Anthropology* 19, No. 2 (June 1978): 225–245.

Bettelheim, Bruno, "Feral children and autistic children," *American Journal of Sociology* 64, No. 5 (March 1959): 455–467.

Bhaumik, Subir, "Tsunami folklore 'saved islanders'," *BBC News*, January 20, 2005. http://news.bbc.co.uk/2/hi/south_asia/4181855.stm, accessed December 31, 2016.

Bickerton, D., *Species and Language* (Chicago: University of Chicago Press, 1990).

Bidelman, Gavin M., Sylvain Moreno, and Claude Alain, "Tracing the emergence of categorical speech perception in the human auditory system," *Neuroimage* 79 (October 1, 2013): 201–212. Epub May 3, 2013, doi: 10.1016/j.neuroimage.2013.04.093.

Biedermann, P.H.W., and M. Taborsky, "Larval helpers and age polyethism in ambrosia beetles," *Proceedings of the National Academy of Sciences, USA* 108 (2011): 17064–17069.

Bielicki, T., "Evolution of the intensity of feedbacks between physical and cultural evolution from man's emergence to present times," UNESCO Expert Meeting on the Biological Aspects of Race, Moscow, August 12–18, 1964: 1–3.

Binford, Lewis, *In Pursuit of the Past: Decoding the Archaeological Record* (Berkeley: University of California Press, 2002).

Binkofski, F., and G. Buccino, "The role of ventral premotor cortex in action execution and action understanding," *Journal of Physiology (Paris)* 99, No. 4–6 (June 2006): 396–405.

Blake, Peter, *The Master Builders* (New York: Norton, 1976).

Blanton, R.E., Gary M. Feinman, Stephen A. Kowalewski, and Peter N. Peregrine, "A dual-processual theory for the evolution of Mesoamerican civilization," *Current Anthropology* 37 (1996): 1–14.

Blinkov, S.M., and I.I. Glezer, *The Human Brain in Figures and Tables: A Quantitative Handbook* (New York: Plenum Press, 1968).

Boakes, R.A., and I. Gaertner, "The development of a simple form of communication," *Quaterly Journal of Experimental Psychology* 29 (1977): 561–575.

Boas, Franz, "Review of the passing of the great race," *American Journal of Physical Anthropology* 1, No. 3 (1918): 363.

Boas, Franz, *Contributions to the Ethnology of the Kwakiutl* (New York: Columbia University Press, 1925).

Boas, Franz, *The Mind of Primitive Man* (New York: Macmillan, 1911).

Boas, Franz, "The methods of ethnology," *American Anthropologist* 22, No. 4 (1920): 311–321.

Boddy, A.M., M.R. McGowen, C.C. Sherwood, L.I. Grossman, M. Goodman, and D.E. Wildman, "Comparative analysis of encephalization in mammals reveals relaxed constraints on anthropoid primates and cetacean brain scaling," *Journal of Evolutionary Biology* 25 (2012): 981–994.

Bogin, Barry, *Patterns of Human Growth* (Cambridge: Cambridge University Press, 1999).

Bogin, B., *The Growth of Humanity* (New York: John Wiley and Sons, 2001).

Bohnet, Iris, *What Works: Gender Equality by Design* (Cambridge: Belknap Press, 2016).

Bond, J.E., E. Roberts, C.G. Woods et al., "ASPM is a major determinant of cerebral cortical size," *Nature Genetics* 32, No. 2 (October 2002): 316–320.

Bonner, Stanley F., *Education in Ancient Rome* (Brussels: Société d'Études Latines de Bruxelles, 1978).

Bonner, John Tyler, *The Evolution of Culture in Animals* (Princeton: Princeton University Press, 1980).

Boss, Leslie P., "Epidemic hysteria: a review of the published literature," *Epidemiologic Reviews* 19, No. 2 (1997): 233–243.

Boserup, Ester, *The Conditions of Agricultural Growth: The Economics of Agrarian Change under Population Pressure* (Chicago: Aldine, 1965).

Bouchard, T.J., Jr., and M. McGue, "Familial studies of intelligence: a review," *Science* 212 (1981): 1055–1059.

Bouchard, T.J., "I.Q. and human intelligence," *Science* 284, No. 5416 (1999): 922–923.

Bourgois, Philippe, *In Search of Respect*, 2nd edition (Cambridge: Cambridge University Press, 2003).

Bourgois, Philippe, and Jeff Schonberg, *Righteous Dopefiend* (Berkeley: University of California Press, 2009).

Boulter, Michael, *Extinction: Evolution and the End of Man* (New York: Columbia University Press, 2002).

Boyd, C.E., *Public Libraries in Ancient Rome* (Chicago: University of Chicago Press, 1915).

Brach, Vincent, "*Anelosimus studiosus* (Araneae: Theridiidae) and the evolution of quasisociality in Theridiid spiders," *Evolution* 31, No. 1 (March 1977): 154–161.

Brafman, Ori, and Rom Brafman, *Sway: The Irresistable Pull of Irrational Behavior* (New York: Doubleday, 2008).

Brain, C.K., and A. Sillent, "Evidence from the Swartkrans cave for the earliest use of fire," *Nature* 336 (December 1, 1988): 464–466.

Brand, Stewart, "What squatter cities can teach us," (2010). www.ted.com/talks/stewart_brand_on_squatter_cities.html, accessed December 31, 2016.

Braun, John, "The 'Speculative Rhetoric' of Charles Sanders Peirce," *Philosophy and Rhetoric* 14, No. 1 (Winter 1981): 1–15.

Brennan, Patricia L.R., Tim R. Birkhead, Kristof Zyskowski, Jessica van der Waag, and Richard O. Prum, "Independent evolutionary reductions in the phallus of basal birds," *Journal of Avian Biology* 39, No. 5 (September 2008): 487–492.

Brewis, Alexandra, Karen Schmidt, and Mary Meyer, "ADHD-type behavior and harmful dysfunction in childhood: a cross-cultural model," *American Anthropologist* 102, No. 4 (December 2000): 823–828.

Brewis, Alexandra, Karen Schmidt, and Mary Meyer, "On the biocultural study of children's hyperactive and inattentive behavior," *American Anthropologist* 104, No. 1 (March 2002): 1–14.

Brewis, Alexandra, and Karen L. Schmidt, "Gender variation in the identification of Mexican Children's psychiatric symptions," *Medical Anthropology Quarterly* 17, No. 3 (2003): 376–393.

Brizendine, Louanne, *The Female Brain* (New York: Random House, 2006).
Broadhurst, C.L., S.C. Cunnane, and M.A. Crawford, "Rift Valley lake fish and shellfish provided brain-specific nutrition for early Homo," *British Journal of Nutrition* 79 (1998): 2–21.
Brodie, E.D., "Caution: Niche construction ahead," *Evolution* 59 (2005): 249–251.
Brody, Nathan, "g and the One–many problem: Is one enough?" Chapter 8 in *The Nature of Intelligence: Novartis Foundation Symposium 233* (2008), edited by Gregory R. Bock, Jamie A. Goode, and Kate Webb. doi: 10.1002/0470870850.
Bromenshenk, Jerry J., Colin B. Henderson, Charles H. Wick et al., "Iridovirus and microsporidian linked to honey bee colony decline," *PLoS ONE* 5, No. 10 (October 2010): 1–11.
Brown, Culum, "Fish intelligence, sentience and ethics," *Animal Cognition* 18, No. 1 (2014): 1–17.
Brown, Donald, *Human Universals* (New York: McGraw-Hill, 1991).
Brown, R., *Words and Things* (New York: Free Press, 1958).
Brugmans, Saartje, Martin P.J. van Boxtel, Eris F.P.M. Vuurman, and Floort Jolles, "The prevalence of cortical gray matter atrophy may be over estimated in healthy aging brain," *Neuropsychology* 23, No. 9 (2009): 541–550.
Bruns, Roger, *Knights of the Road: A Hobo History* (New York: Methuen, 1980).
Buckley, S.J., "Language development in children with Down syndrome – Reasons for optimism," *Down Syndrome Research and Practice* 1, No. 1 (1993): 3–9.
Budiansky, Steven, *If a Lion Could Talk: The Evolution of Language and Consciousness* (New York: Free Press, 1998).
Bullock, Mark A., and David H. Grinspoon, "Global Climate Change on Venus," *Scientific American* 280 (March 1999): 51–57.
Bunn, Henry T., Ellen M. Kroll, Stanley H. Ambrose et al., "Systematic butchery by Plio/Pleistocene hominids at Olduvai Gorge, Tanzania" and Comments and Reply, *Current Anthropology* 27, No. 5 (December 1986): 431–452.
Burda, H., R.L. Honeycutt, S. Begall et al., "Are naked and common mole rats eusocial and if so, why?" *Ecology and Sociobiology* 47 (2000): 293–303.
Burkart, J.M., S.B. Hrdy, and C.P. Van Schaik, "Cooperative breeding and human cognitive evolution," *Evolutionary Anthropology* 18 (2009): 175–186.
Burns, James G., Julien Foucaud, and Frederic Mery, "Costs of memory: lessons from 'mini' brains," *Proceedings of the Royal Society B* 278, No. 1707 (December 22, 2010): 923–929.
Butler, Ann B., and Rodney M.J. Coterill, "Mammalian and avian neuroanatomy and the question of consciousness in birds," *Biological Bulletin* 211, No. 2 (October 2006): 106–127.
Byrne, R.W., and A. Whiten, *Machiavellian Intelligence: Social Expertise and the Evolution of Intellect in Monkeys, Apes and Humans* (Oxford: Oxford Science, 1988).
Cachel, S., "The natural history origin of human intelligence: a new perspective on the origin of human intelligence," *Social Neuroscience Bulletin* 7, No. 1 (1994): 25–30.
Caesar, Gaius Julius (58 B.C.E.) *The Gallic War*, translated by H.J. Edwards (London: Heinemann, 1939 edition).

Cahn, Alvin R., "Records and distribution of the fossil beaver, Castoroides Ohioensis," *Journal of Mammalogy* 13, No. 3 (August 1932): 229–241.

Caldararo, Niccolo, "Lost libraries of the Americas," *Library History* 10 (1994): 9–18.

Caldararo, Niccolo, "The HIV/AIDS epidemic: its evolutionary implications for human ecology with special reference to the immune system," *Science of the Total Environment* 191 (1996): 245–269.

Caldararo, Niccolo, "Comment on Brewis et al.," *American Anthropologist* 104, No. 1 (March 2002a): 282–283.

Caldararo, Niccolo, "Ancient DNA and human origins: the role of gene sequence variation and the species concept," *Human Nature Review* 2 (2002b): 317–321. http://www.human-nature.com/nibbs/02/caldararo.html, accessed January 1, 2017.

Caldararo, Niccolo, "Origins and nature of fundamentalism in society," *Human Nature Review* 2 (October 2002c): 469–482.

Caldararo, Niccolo, "Human ecological intervention and the role of forest fires in human ecology," *Science of the Total Environment* 292 (2002d): 141–165.

Caldararo, Niccolo, "Phylogenetics and ancient DNA the role of gene sequence variation in the species concept," *The Linnean* 19 (2003a): 27–29.

Caldararo, Niccolo, "Japan and the concept of Sustainability," *International Journal of Anthropology* 19, No. 3 (2003b): 213–230.

Caldararo, Niccolo, *Sustainability, Human Ecology and the Collapse of Complex Societies* (Lewiston: The Edwin Mellen Press, 2004a).

Caldararo, Niccolo, "Some effects of the use of ultrasonic devices in conservation and the question of standards for cleaning objects," *Objects Specialty Group Postprints*, American Institute for Conservation of Historic and Artistic Works 11 (2004b): 126–153.

Caldararo, Niccolo, "War, mead and nature of criticism in anthropology," *Anthropological Quarterly* 77, No. 2 (2004c): 311–322.

Caldararo, Niccolo, "The tendency to make man an exception," in *Primatology: Theories, Methods and Research, Animal Science, Issues and Professions Series*, edited by Emil Potocki and Juliusz Krasinski (Hauppauge, NY: Nova Science, 2009a): 113–128.

Caldararo, Niccolo, "Primitive and modern economics: derivatives, liquidity, value, panic and crises, a uniformitarian view," *Forum for Social Economics* 38, No. 1 (2009b): 31–51.

Caldararo, Niccolo, "Caching, money, magic, derivatives, mana and modern finance," *Journal of World Anthropology, Occasional Papers* 3, No. 2 (2009c): 1–47.

Caldararo, Niccolo, "Derivatives and debt: the market as God and marketing as proselytizing," in *Finance and Banking Developments*, edited by Charles V. Karsone (Hauppauge, NY: Nova Science, 2011): 99–117.

Caldararo, Niccolo, "Complex animal societies, domestication and human uniqueness," *Journal of Primatology* 1, No. 1 (2012a). doi: 10.4172/2167-6801.1000103.

Caldararo, Niccolo, "Evolutionary aspects of disease avoidance: the role of disease in the development of complex society" (2012b), https://papers.ssrn.com/sol3/papers.cfm?abstract_id=2001098, accessed January 1, 2017.

Caldararo, Niccolo, *The Psychic Unity of Mankind: The Origins of Anthropology, the Anti-slavery Movement, Cultural Relativism and Man's "Unique" Nature* (Saarbrucken: Scholars' Press, 2014).

Caldararo, Niccolo, *The Anthropology of Complex Economic Systems: Inequality, Stability and Cycles of Crisis* (Lanham, MD: Lexington Press, 2013).

Caldararo, Niccolo, "Beyond zero population: Ethnohistory, archaeology and the Khmer, climate change and the collapse of civilizations," *Anthropology* 3, No. 2 (2015): 1–5.

Caldararo, Niccolo, "Social behaviour and the superorganism: implications for disease and stability in complex animal societies and Colony Collapse Disorder in honeybees," *Interdisciplinary Description of Complex Systems* 13, No. 1 (2015a): 82–98.

Caldararo, Niccolo, "Local resistance in the era of capitalist globalization: clash of cultures in the 21st century," *The IUP Journal of International Relations* IX, No. 3 (2015b): 1–22.

Caldararo, Niccolo, "The origins of the nation-state in Japan: destruction of militant buddhism and western contact of the 16th and 17th centuries: implications for the concept of the state," *American Studies Eurasian Perspective* 1, No. 2 (2016a): 87–98.

Caldararo, Niccolo, "Denisovans, Melanesians, Europeans, and Neandertals: the confusion of DNA assumptions and the biological species concept," *Journal of Molecular Evolution* 83, No. 1–2 (2016b): 78–87.

Caldararo, Niccolo, "Financial evolution and central bank credit: ethics, morals, taxation, stagnation and bank holidays," *International Finance and Banking* 3, No. 2 (2016c): 1–32.

Caldararo, Niccolo, and Steve Gabow, "Mitochondrial DNA analysis and the place of Neandertals in Homo," *Ancient Biomolecules* 3 (2000): 135–158.

Caldararo, Niccolo, and Michael Guthrie, "Mitochondrial DNA, the Y chromosome and the origins of modern humans," *Homo* 49, No. 2 (1998): 225–240.

Caldararo, Niccolo, and Michael Guthrie, "A note on the Denisova Cave mtDNA sequence," *Nature Proceedings* 2 (April 2012): 1–13.

Caldararo, Niccolo, and Martha Senger, "The longest rent strike: the occupation and self-government of an apartment building in San Francisco," Manuscript (1979–2012).

Caldwell, Roy L., Richard Ross, Arcadio Rodaniche, Christine L. Huffard, "Behavior and body patterns of the larger Pacific striped octopus," *PLoS ONE* 10, No. 8 (2015): e0134152. doi: 10.1371/journal.pone.0134152.

Camerer, Colin F., "Strategizing in the brain," *Science* 300 (June 13, 2003): 1673–1675.

Campbell, Anne, *A Mind of Her Own: The Evolutionary Psychology of Women* (Oxford: Oxford University Press, 2002).

Campbell, D., "Downward causation in hierarchically organized biological systems," in *Studies in the Philosophy of Biology: Reduction and Related Problems*, edited by F. Ayala and T. Dobzhansky (Basingstoke: Palgrave, 1974): 179–186.

Campbell, D.T., "Legal and primary – group social controls," in *Law, Biology and Culture,* edited by Margaret Gruter and Paul Bohannon (Santa Barbara, CA: Ross-Erikson, 1983): 159–170.

Campbell, Joseph, *Hero with a Thousand Faces* (New York: Pantheon Books, 1949).

Campbell, Ruairidh D., Frank Rosell, Bart A. Nolet et al., "Territory and group sizes in Eurasian beavers (Castor fiber): Echoes of settlement and reproduction?" *Behavioral Ecology and Sociobiology* 58, No. 6 (October 2005): 597–607.

Canetti, Elias, *Crowds and Power* (New York: Viking Press, 1966, 1st edition 1960).

Cangialosi, Karen R., "Social spider defense against kleptoparasitism," *Behavioral Ecology and Sociobiology* 27, No. 1 (1990): 49–54.

Caplan, Arthur L., editor, *The Sociobiology Debate* (New York: Harper, 1978).

Capra, J.A., G.D. Erwin, G. McKinsey, J.L. Rubenstein, and K.S. Pollard, "Many human accelerated regions are developmental enhancers," *Philosophical Transactions of the Royal Society of London B: Biological Sciences* 368, No. 1632 (November 11, 2013).

Caro, Robert, *The Power Broker: Robert Moses and the Fall of New York* (New York: Knopf, 1974).

Carpendale, Jeremy I.M., and Michael Frayn, "Autonomy in ants and humans," *Behavioral and Brain Sciences* 39 (2016): 18–19.

Carrier, David R., "The energetic paradox of human running and hominid evolution," *Current Anthropology* 25, No. 4 (1984): 483–495.

Carter, Richard, and Kamini N. Mendis, "Evolutionary and historical aspects of the burden of malaria," *Clinical Microbiology Reviews* 15, No. 4 (October 2002): 564–594. doi: 10.1128/CMR.15.4.564–594.2002.

Cartmill, Mat, "Human uniqueness and theoretical content in paleoanthropology," *International Journal of Primatology* 11 (1990): 173–192.

Cartmill M., L S B Leakey Lecture, California Academy of Sciences/L S B Leakey Foundation: *Language and Human Evolution* (1999).

Cashdan, Elisabeth, "Ethnic diversity and its environmental determinants: effects of climate, pathogens and habitat diversity," *American Anthropologist* 103, No. 4 (December 2001): 968–991.

Castellanos, F. Xavier, Patti P. Lee, Wendy Sharp et al., "Developmental trajectories of brain volume abnormalities in children and adolescents with attention-deficit/hyperactivity disorder," *JAMA* 288, No. 14 (October 9, 2002): 1740–1748.

Castle, Lon, Ronald E. Aubert, Robert R. Verbrugge, Mona Khalid, and Robert S. Epstein, "Trends in medication treatment for ADHD," *Journal of Attention Disorders* 10, No. 4 (2007): 335–342.

Caspari, R., and Sang-Hee Lee, "Older age becomes common late in human evolution," *Proceedings of the National Academy of Sciences, USA* 101 (2004): 10895–10900.

Cazden, Courtney B., "Review of the language instinct," *American Anthropologist* 97, No. 2 (1995): 375.

Centers for Disease Control and Prevention, "State-based prevalence data of parent reported ADHD medication treatment," CDC 24/7: Saving Lives, Protecting People (2007).

Cezilly, Frank, Marina Préault, Frédérique Dubois, Bruno Faivre, and Bruno Patris, "Pair-bonding in birds and the active role of females: a critical review of the empirical evidence," *Behavioural Processes* 51, No. 1–3 (October 5, 2000): 83–92.

Chang, Edward F., Jochem Rieger, Keith Johnson et al., "Categorical speech representation in human superior temporal gyrus," *Nature Neuroscience* 13 (2010): 1428–1432. doi: 10.1038/nn.2641.

Chang, Ha-Joon, *Bad Samaritans* (New York: Bloomsbury, 2008).

Chapin, F. Stuart, Erika S. Zavaleta, Valerie T. Eviner et al., "Consequences of changing biodiversity," *Nature* 405 (May 11, 2000): 234–242.

Chapple, Eliot Dismore, and Conrad Arensburg, *Measuring Human Relations* (Provincetown, MA: The Journal Press, 1940).

Chauvin, Remy, *Animal Societies: From the Bee to the Gorilla* (New York: Hill and Wang, 1968).

Chenn, A., and C.A. Walsh, "Regulation of cerebral cortical size by control of cell cycle exit in neural precursors," *Science* 297, No. 5580 (July 19, 2002): 365–369.

Chew, Sing C., *The Recurring Dark Ages* (Lanham, MD: Altamira Press, 2007).

Chew, Sing C., *World Ecological Degradation: Accumulation, Urbanization and Deforestation, 3,000 B.C.–2000 A.D.* (Lanham, MD: Altamira Press, 2001).

Chinawa, J.M., Odetunde, O.I., Obu, Herbert A., et al. "Attention deficit hyperactivity disorder: A neglected issue in the developing world," *Behavioural Neurology* 2014 (2014): 1–6. doi: http://dx.doi.org/10.1155/2014/694764.

Chiu Hsin-hui, *The Colonial 'Civilizing Process' in Dutch Formosa* (Leiden, Netherlands: Brill, 2008).

Chomsky, Noam, *Language and Mind*, enlarged edition (New York: Harcourt Brace and Jovanovich, 1972).

Chomsky, Noam, *Rules and Representations* (New York: Columbia University Press, 1980).

Chomsky, Noam, *The Minimalist Program* (Cambridge: MIT Press, 1995).

Chomsky, Noam, *Cartesian Linguistics: A Chapter in the History of Rationalist Thought* (New York: Harper and Rowe, 1966).

Christakis, D.A., F.J. Zimmerman, D.I. DiGiuseppe et al., "Early television exposure and subsequent attentional problems in children," *Pediatrics* 113, No. 4 (April 2004): 708–713.

Cipolla, Carlo, *Guns, Sails and Empires: Technological Innovation and European Expansion* (New York: Barnes and Noble, 1965).

Cipolla, Carlo, "Editor's introduction," in *The Economic Decline of Empires*, edited by Carlo Cipolla (London: Methuen, 1970): 1–15.

Cipolla, Carlo, *The Basic Laws of Stupidity* (Berkeley: The Mission, 1976).

Cipolla, Carlo, *Clocks and Culture, 1300–1700* (New York: Norton, 1978).

Clark, Andy, "An embodied cognitive science," *Trends in Cognitive Science* 3, No. 9 (1999): 345–351.

Clark, J. Desmond, *The Prehistory of Africa* (New York: Praeger, 1970).

Clifford, James, *The Predicament of Culture* (Cambridge: Harvard University Press, 1988).

Cochran, Gregory, and Henry Harpending, *The 10,000 Year Explosion: How Civilization Accelerated Human Evolution* (New York: Basic Books, 2009).

Coggin, P., "Tossing a coin could well be as insightful to investors as a fund manager," *Financial Times*, July 1, 2006.

Cohn, Mark Nathan, *The Food Crisis in Prehistory* (New Haven: Yale University Press, 1977).

Coles, J.M., and E.S. Higgs, *The Archaeology of Early Man* (New York: Faber and Faber, 1969).

Collias, Nicholas E., and Elsie C. Collias, *Nest Building and Bird Behavior* (Princeton: Princeton University Press, 1984).

Connolly, Cornelius J., *External Morphology of the Primate Brain* (Springfield: Charles C. Thomas, 1950).

Connor, Richard C., "Dolphin social intelligence: complex alliance relationships in Bottlenose dolphins and a consideration of selective environments for extreme brain size evolution in mammals," *Philosophical Transactions: Biological Sciences* 362, No. 1480 (2007): 587–602.

Conrad, B., and S.E. Antonarakis, "Gene duplication: a drive for phenotypic diversity and cause of human disease," *Annual Review of Genomics and Human Genetics* 8 (2007): 17–35.

Conrad, Donald, Dalila Pinto, Richard Redon et al., "Origins and functional inpact of copy number variation in the human genome," *Nature* 464 (April 2010): 704–712.

Cook, Edwin H., Jr., and Stephen W. Scherer, "Copy-number variations associated with neuropsychiatric conditions," *Nature* 455 (October 2009): 919–923.

Cook, Tim, *Vagrancy: Some New Perspectives* (New York: Academic Press, 1979).

Conroy, G.C., *Primate Evolution* (New York: W.W. Norton, 1990).

Cordain, Loren, *The Paleo Diet: Lose Weight and Get Healthy* (Hoboken: John Wiley and Sons, 2002).

Corkery, Michael, "Wells Fargo fined $185 million for fraudulently opening accounts," *New York Times*, September 8, 2016.

Cornelio, Alianda, Ruben E. de Bittencourt-Navarrete, Ricardo de Bittencourt Brum et al., "Human brain expansion during evolution is independent of fire control and cooking," *Frontiers in Neuroscience* 10 (April 10, 2016): 167.

Cornell, H.N., J.M. Marzluff, and S. Pecararo, "Social learning spreads knowledge about dangerous humans among American crows," *Proceedings of the Royal Society* 279 (2012): 499–508.

Courage, Katherine Harmon, "How the freaky octopus can help us understand the human brain," *National Geographic Magazine*, October 1, 2013.

Courage, Katherine Harmon, "Social octopus species shatters beliefs about ocean dwellers: group-living animals are helping scientists learn more about octopus behavior," *National Geographic Magazine*, July 30, 2014.

Coulton, G.G., *Medieval Village, Manor and Monastery* (New York: Harper Torchbooks, 1925).

Count, Earl W., "Brain and body weight in man: their antecedents in growth and evolution. A study in dynamic somatometrics," *Annals of the New York Academy of Sciences* XLVI, Article 10 (1947): 993–1122.

Cowan, Nelson, "The magical number 4 in short-term memory: a reconsideration of mental storage capacity," *Behavioral and Brain Sciences* 24, No. 1 (2001): 87–114.

Crabtree, Gerald R., "Our fragile intellect," Parts I & II, *Trends in Genetics* 29, No. 1 (January 2013).

Crabtree, James, "Resourceful residents address Mumbai's housing problem," *Financial Times*, April 1, 2012, 4.

Crail, T., *Apetalk and Whalespeak* (Los Angeles: J. P. Tarcher, 1981).

Crapo, Richley, *Anthropology of Religion* (Columbus: McGraw-Hill, 2002).

Creel, Scott, David W. Wildt, and Steven Monfort, "Aggression, reproduction, and androgens in wild dwarf mongooses: a test of the challenge hypothesis," *The American Naturalist* 141, No. 5 (May 1993): 816–825.

Creutzfeldt, O.D., and Mary Creutzfeldt, "The allocortex and limbic system," Chapter 9 in *Cortex Cerebri: Performance, Structure and Functional Organization of the Cortex* (Oxford: Oxford University Press, 1995): 487–540.

Crick, Francis, *The Astonishing Hypothesis: The Scientific Search for the Soul* (New York: Charles Scribner's Sons, reprint edition, 1995).

Crisp, Alastair, Chaira Boschetti, Malcolm Perry et al., "Expression of multiple horizontally acquired genes is a hallmark of both vertebrate and invertebrate genomes," *Genome Biology* 16, No. 50 (2015). doi: 10.1186/s13059-015-0607-3.

Croner, R.F., "The basis of childhood dysphasia: a linguistic approach," in *Development Dysphasia*, edited by M.A. Wyke (London: Academic Press, 1978): 85–134.

Crowder, R.G., *Principles of Learning and Memory* (Hillsdale, NJ: Lawrence Erlbaum, 1976).

Crumley, Carole L., "Heterarchy and the analysis of complex societies", introduction to *Archaeological Papers of the American Anthropological Association 6*, "Heterarchy and the Analysis of Complex Societies," edited by R.M. Ehrenreich, Carole L. Crumley, and Janet E. Levy (Washington, DC: The American Anthropological Association, 1995): 1–6.

Cucchiaro, Matt Aames, "On the myth of Ape language: Noam Chomsky interviewed," (2007). https://chomsky.info/2007____/, accessed January 2, 2017.

Cummins, R., "The LOT of the causal theory of mental content," *Journal of Philosophy* XCIV, No. 10 (October 1997): 535–542.

Cutler, R., "Evolution of human longevity and the genetic complexity governing aging rate," *Proceedings of the National Academy of Sciences, USA* 72 (1975): 664–668.

Dahlberg, Frances, "Introduction," in *Woman the Gatherer*, edited by Frances Dahlberg (New Haven: Yale University Press, 1981): 1–24.

Dalton, Doug, "Cargo cults and discursive madness," *Oceania* 70, No. 4, "A Critical Retrospective on 'Cargo Cult': Western/Melanesian Intersections" (June 2000): 345–361.

Damasio, A.R., "The somatic marker hypothesis and the possible functions of the prefrontal cortex," *Philosophical Transactions of the Royal Society B: Biological Sciences* 351, No. 1346 (1996): 1413–1420.

Damasio, A.R., *The Feeling of What Happens: Body and Emotion in the Making of Consciousness* (New York: Harcourt, 1999).

Damilov, P.I. and Fyodorov, F.V., "Characterization of the building activity of Canadian and European beavers and Northern European Russian," *Russian Journal of Ecology* 46, No. 3 (2015): 273–278.

Damuth, John, "A macroevolutionary explanation for energy equivalence in the scaling of body size and population density," *The American Naturalist* 169, No. 3 (2007): 621–631.
Danforth, B.N., "Evolution of sociality in a primitively eusocial lineage of bees," *Proceedings of the National Academy of Sciences, USA* 99 (January 8, 2002): 286–290.
Daniau, Anne-Laure, Francesco d'Errico, and Maria Fernanda Sánchez Goñi, "Testing the hypothesis of fire use for ecosystem management by Neanderthal and Upper Palaeolithic modern human populations," *PLoS ONE* 5, No. 2 (2010), e9157: 1–10.
Danielson, Karen A., and Robert E. Lang, "The case for higher density housing: a key to smart growth?" in *ULI on the Future: Smart Growth: Economy, Community, and Environment* (Charlotte, NC: The Metropolitan Institute at Virginia Tech, 1998): 1–8.
Damilov, P.I. and Fyodorov, F.V., "Characterization of the building activity of Canadian and European Beavers and Northern European Russian," *Russian Journal of Ecology* 46, No. 3 (2015): 273–278.
Dantzer, R., J.C. O'Connor, Gregory G. Freund, R.W. Johnson, and Keith W. Kelley, "From inflammation to sickness and depression: when the immune system subjugates the brain," *Nature Neuroscience* 9 (January 2008): 46–57.
Darlington, Cynthia L., *The Female Brain (Conceptual Advances in Brain Research)*, 2nd edition (Boca Raton: CRC Press, 2009).
Dart, R., "The predatory implemental technique of Australopithecus," *American Journal of Physical Anthropology* 7 (1949): 11–38.
Dart, R., "The predatory transition from ape to man," *International Anthropological and Linguistic Review* 1 (1954): 201–213.
Darwin, Charles, *The Descent of Man*, 2nd edition (Chicago: W.B. Conkey, 1874). Original work published in 1871.
Darwin, C. *The Autobiography of Charles Darwin* (New York: W.W. Norton, 1969 edition).
Davis, Mike, *Planet of Slums* (London: Verso Press, 2006).
Dawkins, Richard, *The Extended Phenotype* (Oxford: Oxford University Press, 1982).
Dawkins, R., *The Selfish Gene* (Oxford: Oxford University Press, 1989). Originally published in 1976.
Dawkins, R., "Extended phenotype – but not too extended. A reply to Laland, Turner and Jablonka" *Biology and Physiology* 19 (2004): 377–396.
Dawson, Warren A., "Education in Ancient Egypt," *Science Progress in the Twentieth Century (1919–1933)* 20, No. 77 (July 1925): 109–119.
Day, Michael H., *Guide to Fossil Man*, 4th edition (Chicago: University of Chicago Press, 1986).
Deacon, Terrence W., *The Symbolic Species: The Co-evolution of Language and Brain* (New York: W.W. Norton, 1997).
Deaner, Robert O., Carel P. van Schaik, and Valen Johnson, "Do some taxa have better domain-general cognition than others? A meta-analysis of non-human primate studies," *Evolutionary Psychology* 4 (2006): 149–196.
Deaner, Robert O., Karin Isler, Judith Burkart, and Carel van Schaik, "Overall brain size, and not encephalization quotient, best predicts cognitive ability across non-human primates," *Brain, Behavior and Evolution* 70, No. 2 (2007): 115–124.
Deary, I.J., L. Penke, *and* W. Johnson, "The neuroscience of human intelligence differences," *Nature Reviews Neuroscience* 11 (2010): 201–211.

Dennis, M.Y., X. Nuttle, P.H. Sudmant et al., "Evolution of human-specific neural SRGAP2 genes by incomplete segmental duplication," *Cell* 149 (2012): 912–922.

Denton, D.A., M.J. Mckinley, and R.S. Weisinger, "Hypothalamic integration of body fluid regulation," *Proceedings of the National Academy of Sciences, USA* No. 14 (July 9, 1996): 7397–7404.

D'Errico, Francesco, "The invisible frontier: a multiple species model for the origin of behavioral modernity," *Evolutionary Anthropology* 12, No. 2 (2003): 188–202.

d'Errico, F., and A. Nowell, "A new look at the Berekhat Ram figurine: implications for the origins of symbolism," *Cambridge Archaeological Journal* 10 (2000): 123–167.

DeFelipe, Javier, Lidia Alonso-Nanclares, and Jon Arellano, "Microstructure of the neocortex: comparative aspects," *Journal of Neurocytology* 31 (2002): 299–316.

De Ruiter, Jan, Gavin Weston, and Stephen M. Lyon, "Dunbar's number: group size and brain physiology in humans reexamined," *American Anthropologist* 133, No. 4 (2011): 557–568.

Department of the Environment, Food and Rural Affairs, "Key facts about waste and recycling: construction and demolition waste management: 1999–2005," *Digest of Environmental Statistics* (UK: DEFRA, 2006).

Desert, Michel, Marie Preaux, and Robin Tund, "So young and already victims of stereotype threat: socio-economic status and performance of 6 to 9 years old children on Raven's progressive matrices," *European Journal of Psychology of Education* 24, No. 2 (June 2009): 207–218.

DeSilva, Jeremy M., and Julie J. Lesnik, "Brain size at birth throughout human evolution: a new method for estimating neonatal brain size in hominins," *Journal of Human Evolution* 55, No. 6 (December 2008): 1064–1074.

De Smet, H.J., P. Paquier, J. Verhoeven, and P. Marien, "The cerebellum: its role in language and related cognitive and affective functions," *Brain and Language* 127, No. 3 (2013): 334–342. doi: 10.1016/j.bandl.2012.11.001.

Detrain, C., J.L. Deneubourg, and J.M. Pasteels, *Information Processing in Social Insects* (New York: Harvard University Press, 1999).

Devine, J.V., "Stimulus attributes and training procedures in learning-set formation of rhesus and cebus monkeys," *Journal of Comparative and Physiological Psychology* 73, No. 1 (1970): 62–67.

De Waal Malefijt, Annemarie, *Religion and Culture* (Prospect Heights: Waveland Press, 1968).

De Waal, Frans, *Are We Smart Enough to Know How Smart Animals Are?* (New York: W.W. Norton, 2016).

Diamond, Jared, *Guns, Germs and Steel* (New York: W.W. Norton, 1997).

Diamond, Jared, *Collapse* (New York: Viking Press, 2005).

Diamond, Marian Cleeves, *Enriching Heredity: The Impact of the Environment on the Anatomy of the Brain* (New York: The Free Press, 1988).

Dilthey, Wilhelm, "The construction of the historical world in the human sciences," in *Dilthey Selected Writings*, edited by H.P. Rickman (Cambridge: Cambridge University Press, 1914): 168–245.

Diop, Cheikh Anta, *The African Origin of Civilization: Myth or Reality?* edited and translated by Mercer Cook (Chicago: Lawrence Hills Books, 1974).

Dissing-Olesen, Lasse, Jeffrey M. LeDue, Ravi L. Rungta, Jasmin K. Hefendehl, Hyun B. Choi, and Brian A. MacVicar, "Activation of neuronal NMDA receptors triggers transient ATP-mediated microglial process outgrowth," *Journal of Neuroscience* 34, No. 32 (August 6, 2014): 10511–10527.

Doan, R.N., B.L. Bae, B. Cubelos et al., "Mutations in human accelerated regions disrupt cognition and social behavior," *Cell* 167, No. 2 (October 6, 2016): 341–354.

Dolhinow, Phyllis Jay, "The ecology and social behavior of the Indian langur monkey" (Ph.D. diss., University of Chicago, 1963).

Dolhinow, Phyllis Jay, "The common langur ofNorth India," in *Primate Behavior: Field Studies of Monkeys and Apes*, edited by Irven DeVore (New York: Holt, Rinehart and Winston, 1965): 197–249.

Dombey, Daniel, "Istanbul boom belies slower growth," *Financial Times*, Tuesday, April 3, 2012: 2.

Donald, Merlin, *A Mind So Rare* (New York: W.W. Norton, 2002).

Donovan, John, and Caren Zucker, *In a Different Key: The Story of Autism* (New York: Crown Press, 2016).

Doolittle, W. Ford, "Phylogenetic classification and the universal tree," *Science* 284, No. 5423 (June 25, 1999): 2124–2128.

Douglas, Mary, *Purity and Danger: An Analysis of Concepts of Pollution and Taboo* (New York: Praeger, 1966).

Dow, Gregory K., and Clyde G. Reed, "The origins of sedentism: climate, population and technology," *Journal of Economic Behavior and Organization* 119 (November 2015): 56–71.

Dubal, Dena B., Jennifer S. Yokoyama, Lei Zhu et al., "Life extension factor Klotho enhances cognition," *Cell Reports* 7, No. 4 (May 22, 2014): 1065–1076.

Duchaine, Bradley, Leda Cosmides, and John Tooby, "Evolutionary psychology and the brain," *Current Opinion in Neurobiology* 11 (2001): 225–230.

Duesenberg, David B., *Living at Micro Scale* (Cambridge: Harvard University Press, 2009).

Duffles, Marilia, "Hawaii looks back for the future," *Financial Times*, April 8, 2006.

Duffy, J. Emmett, "Eusociality in a coral-reef shrimp," *Nature* 381 (June 6, 1996): 512–514.

Duffy, J. Emmett, "The ecology and evolution of eusociality in sponge-dwelling shrimp," in *Genes, Behavior, and Evolution in Social Insects*, edited by T. Kikuchi (Sapporo, Japan: University of Hokkaido Press, 2002): 1–38.

Duffy, J. Emmett, and Kenneth S. Macdonald, "Colony structure of the social snapping shrimp *Synalpheus filidigitus* in Belize," *Journal of Crustacean Biology* 19, No. 2 (1999): 283–292.

Dumont, Luis, *Homo Hierarchicus: The Caste System and Its Implications (Nature of Human Society)* (Chicago: University of Chicago Press, 1966).

Dunbar, R.I.M., "Neocortex size as a constraint on group size in primates," *Journal of Human Evolution* 22, No. 6 (1992): 469–493.

Dunbar, Robin, "Coevolution of neocortical size, group size and language in humans," *Behavioral and Brain Sciences* 16, No. 4 (1993): 681–735.

Dunbar, R.I.M., *Grooming, Gossip and the Evolution of Language* (New York: Faber and Faber, 1996).

Dunbar, R.I.M., "The social brain hypothesis," *Evolutionary Anthropology* 6, No. 5 (1998): 178–190.

Dunbar, R.I.M., Chris Knight, and Camilla Power, *The Evolution of Culture: A Historical and Scientific View* (New Brunswick: Rutgers University Press, 1999).

Dunbar, R.I.M., and Susanne Shultz, "Evolution of the social brain," *Science* 317, No. 5843 (2007): 1344–1347.

Dunbar, R.I.M., and Louise Barrett, editors, *Oxford Handbook of Evolutionary Psychology* (Oxford: Oxford University Press, 2007).

Dunbar, R.I.M., *How Many Friends Does One Person Need? "Dunbar's Number" and Other Evolutionary Quirks* (London: Faber and Faber, 2010).

Dunbar, R.I.M., "How conversations around campfires came to be," *Proceedings of the National Academy of Sciences, USA* 111, No. 39 (September 30, 2014): 14013–14014.

Durkheim, E., *The Division of Labor in Society*, translated by G. Simpson (New York: Macmillan, 1933). Originally published in 1893.

Durkheim, E., *The Elementary Forms of the Religious Life* (New York: Free Press, 1965). Original work published in 1915.

Dwivedi, K.N., and R.G. Banhatti, "Attention deficit/hyperactivity disorder and ethnicity," *Archives of Disease in Childhood* 90, Suppl. 1 (2005): 10–12.

Dworkin, M., "The Myxobacteria: new directions in studies of procaryotic development," *Critical Reviews in Microbiology* 2 (1972): 435–452.

Dyer, Fred C., and Thomas D. Seeley, "Dance dialects and foraging range in three Asian honey bee species," *Behavioral Ecology and Sociobiology* 28 (1991): 227–233.

Eckmeier, Eileen, Reuven Yeshurun, Mina Weinstein-Evron, Eugenia Mintz, and Elisabetta Boaretto, "14C Dating of the Early Natufian at el-Wad Terrace, Mount Carmel, Israel: methodology and materials characterization," *UAiR Radiocarbon* 54, No. 3–4 (2012): 823–836.

Edelman, Gerald, *Neural Darwinism. The Theory of Neuronal Group Selection* (New York: Basic Books, 1987).

Editorial, *The Economist*, March 31, 2007: 58.

Edwards, Stephen W., and Richard W. Clinnick, "Keeping the Lower Palaeolithic in perspective," *Man*, New Series, 15, No. 2 (June 1980): 381–383.

Ehmann, R., E. Boedeker, U. Friedrich et al., "Canine scent detection in the diagnosis of lung cancer: revisiting a puzzling phenomenon," *European Respiratory Journal,* 39 (January 1, 2011): 669–676. doi: 10.1183/09031936.00051711.

Eiseley, L., *The Immense Journey* (New York: Random House, 1958).

Eisenberg, Dan T.A., Benjamin Campbell, Peter Gray, and Michael D. Sorenson, "Dopamine receptor genetic polymorphisms and body composition undernourished pastoralists: an exploration of nutrition indices among nomadic and recently settled Ariaal men of Northern Kenya," *BMC Evolutionary Biology* 8, No. 173 (2008). doi: 10.1186/1471-2148-8-173.

Eisert, Regina, Charles W. Potter, and Olav T. Oftedal, "Brain size in neonatal and adult Weddell seals: costs and consequences of having a large brain," *Marine Mammal Science* 30, No. 1 (January 2013): 184–205.

Ekholm, Gordon F., "Wheeled toys in Mexico," *American Anthropologist* 11, No. 4 (1946): 222–228.

Ekirch, A. Roger, *At Day's Close: Night in Times Past* (New York: W.W. Norton, 2005).

El Albani, Abderrazak, Stefan Bengtson, Donald Eugene Canfield et al., "Large colonial organisms with coordinated growth in oxygenated environments 2.1 Gyr ago," *Nature* 466, No. 7302 (July 2010): 100–104.

Elbroch, Mark, *Animal Skulls: A Guide to North American Species* (Mechanicsburg, PA: Stackpole Books, 2006).

Elbroch, L.M., Patrick E. Lendrum, Howard Quigley, and Anthony Caragliuo, "Spatial overlap in a solitary carnivore: support for the land tenure, kinship or resource dispersion hypotheses?" *Journal of Animal Ecology* 85, No. 2 (October 16, 2015): 487–496.

Elia, Josephine, Joseph T. Glessner, Kai Wang et al., "Genome-wide copy number variation study associates metabolic glutamate receptor gene networks with attention deficit hyperactivity disorder," *Nature Genetics* 44 (June 24, 2012): 78–84.

Eliade, M., *Shamanism: Archaic Techniques of Ecstasy*, Bolligen Series 76 (New York: Pantheon, 1964).

Elias, Norbert, *The Civilizing Process* (Oxford: Basil Blackwell, 1978). Originally published in 1939.

Eller, Cynthia, *The Myth of Matriarchal Prehistory: Why an Invented Past Won't Give Women a Future* (Boston: Beacon Press, 2000).

Elman, Jeffrey, *Rethinking Innateness: A Connectionist Perspective on Development* (Cambridge: MIT Press, 1996).

Emery, Nathan J., "Cognitive ornithology: the evolution of Avian intelligence," *Philosophical Transactions of the Royal Society B: Biological Sciences* 361, No. 1465, Reviews (January 29, 2006): 23–43.

Emery, Nathan, Nicola Clayton, and Christopher Frith, *Social Intelligence: From Brain to Culture* (Oxford: Oxford University Press, 2008): 587–602.

Engel, Michael S., Phillip Barden, Mark L. Riccio, and David A. Grimaldi, "Morphologically specialized termite castes and advanced sociality in the Early Cretaceous," *Current Biology* 26 (2016): 522–530.

Eskildsen, Stephen, *Asceticism in Early Taoist Religion* (Albany: State University of New York Press, 1998).

Etzioni, Amitai, *The Moral Dimension: Toward a New Economics* (New York: The Free Press, 1988).

Etzioni, Amitai, *A Comparative Analysis of Complex Organizations*, revised edition (New York: The Free Press, 1975).

Fabrega, Horacio, Jr., *Disease and Social Behavior: An Interdisciplinary Perspective* (Cambridge: MIT Press, 1974).

Fabrega, Horacio, Jr., *Evolution of Sickness and Healing* (Berkeley: University of California Press, 1997, paperback edition).

Fagan, Kevin, "Off the streets into uncertainty," *San Francisco Chronicle,* August 4, 2013.

Fagen, R., *Animal Play Behavior* (New York: Oxford University Press, 1981).

Falk, Dean, "Brain evolution in Homo: The 'radiator' theory," *Behavioral and Brain Sciences* 13, No. 2 (June 1990): 333–344.

Falk, Dean, and B. Dudek, "Mosaic evolution in the neocortex," *Behavioral and Brain Sciences* 16, No. 4 (1993): 701–702.

Falk, D., C. Hildebolt, K. Smith et al., "The brain of LB1, *Homo floresiensis*," *Science* 308, No. 5719 (April 8, 2005): 242–245.

Falk, Dean, "Evolution and the brain," in *Handbook of Paleoanthropology, Primate Evolution and Human Origins*, edited by W. Henke, H. Routhe, and I. Tattersall (Berlin: Springer-Verlag, 2007a): 1–28.

Falk, Dean, "Evolution of the Primate brain," in *Handbook of Paleoanthropology, Primate Evolution and Human Origins*, v. 2 edited by W. Henke, H. Routhe, and I. Tattersall (Berlin: Springer-Verlag, 2007b): 1133–1162.

Falk, D., C. Hildebolt, K. Smith et al., "Brain shape in human microcephalics and *Homo floresiensis*," *Proceedings of the National Academy of Sciences, USA* 104, No. 7 (February 2, 2007): 2513–2518.

Falk, Dean, "Brain lateralization in primates and its evolution in hominids," *Yearbook of Physical Anthropology* 30 (1987): 107–125.

Fanon, Franz, *Black Skins, White Masks* (New York: Grove Press, 1967).

Fanon, Franz, *The Wretched of the Earth* (London: MacGibbon and Kee, 1961).

Faraone, Stephen V., Joseph Sergeant, Christopher Gillberg, and Joseph Biederman, "The worldwide prevalence of ADHD: is it an American condition?" *World Psychiatry* 2, No. 2 (June 2003): 104–113.

Farris, H.E., and S.E. Breuning, "Post-conditioning habituation and classically conditioned head withdrawal in the red-eared turtle (*Chrysemys scripta elegans*)," *Psychological Record* 27 (1977): 307–313.

Farris, Sarah M., and Susanne Schulmeister, "Parasitoidism, not sociality, is associated with the evolution of elaborate mushroom bodies in the brains of hymenopteran insects," *Proceedings of the Royal Society B* 27 (February 8, 2011): 940–951. doi: 10.1098/rspb.2010.2161.

Farris, S., "Evolution of complex higher brain centers and behaviors: behavioral correlates of Mushroom body elaboration in insects," in *Evolution of Brain Complexity and Animal Minds: 24th Annual Karger Workshop in Evolutionary Neuroscience, New Orleans, LA, October 2012, Special Topic Issue: Brain, Behavior and Evolution 2013, Vol. 82, No. 1,* edited by Leo S. Demski (Basel: S. Karger AG, 2013): 9–18.

Farris, S., and Joseph Van Dyke, "Evolution and function of the insect Mushroom bodies: contributions from comparative and model systems studies," *Current Opinion in Insect Science* 12 (2015): 19025.

Falkowski, Paul G., Miriam E. Katz, Allen J. Milligan et al., "The rise of oxygen over the past 205 million years and the evolution of large placental mammals," *Science* 309, No. 30 (September 2005): 2202–2204.

Faulkes, C.G., and N.C. Bennett, "Family values: group dynamics and social control of reproduction in African mole-rats," *Trends in Ecology and Evolution* 16 (2001): 184–190.

Fazal, Abul, *The Akbar NamA of Abu-l-Fazl*, translated from the Persian by H. Berveridge, 3 vols. (Delhi: Low Price, 1993).

Ferber, Dan, "The education of T-cells," *Science* 316 (April 2007): 191–193.

Ferguson, Brian R., "Tribal warfare," *Scientific American* 266 (January 1992): 108–113.

Ferguson, Niall, *Civilization: The West and the Rest* (London: Penguin Books, 2011).
Ferguson, Niall, *Empire* (New York: Allen Lane, 2002).
Fernando, C., K.K. Karishma, and E. Szathmáry, "Copying and evolution of neuronal topology," *PLoS ONE* 3, No. 11 (2008): 3775.
Fernando, C., "From blickets to synapses: inferring temporal causal networks by observation," *Cognitive Science* 37, No. 8 (2013): 1426–1470.
Ferretti, Francesco, and Jerry A. Fodor, *Mente e Linguaggio* (Rome: Editori Laterza, 2001).
Fialkowski, Konrad R., "A mechanism for the origin of the human brain: a hypothesis," *Current Anthropology* 27, No. 3 (June 1986): 288–290.
Finarelli, John A., and John J. Flynn, "Brain-size evolution and sociality in Carnivora," *Proceedings of the National Academy of Sciences, USA* 106, No. 23 (2009): 9345–9349.
Finlay, B.L., and R.B. Darlington, "Linked regularities in the development and evolution of mammalian brains," *Science* 268 (1995): 1578–1584.
Finlayson, B., S.J. Mithen, M. Najjar et al., "Architecture, sedentism, and social complexity at Pre-Pottery Neolithic A WF16, Southern Jordan," *Proceedings of the National Academy of Sciences, USA* 108, No. 20 (2011): 8183–8188.
Finn, J. K., T. Tregenza, and M.D. Norman, "Defensive tool use in a coconut-carrying octopus," *Current Biology* 19, No. 23 (2009): R1069–R1070.
Firestone, Shulamith, *The Dialectic of Sex* (New York: Bantam Books, 1970).
Firmin, Antenor, *The Equality of the Human Races*, translated by Asselin Charles (Urbana: University of Illinois Press edition, 2002). Originally published in 1885.
Fisher, J., "Categorical perception in animals," in *Encyclopedia of Language and Linguistics*, 2nd edition, edited by Keith Brown (Amsterdam: Elsevier, 2006): 248–251.
Flashman, L.A., N.C. Andresson, M. Flaum, and V.W. Swayze, "Intelligence and regional brain volumes in normal controls," *Intelligence* 25 (1998): 149–160.
Fleissner, Andre, Anna R. Simonin, and N. Louise Glass, "Cell fusion in the filamentous fungus, *Neurospora crassa*," in *Methods in Molecular Biology, vol. 475: Overviews and Methods*, edited by E.H. Chen (Tolowa: Humana Press, 2008): 21–38.
Fluehr-Lobban, Carolyn, "Antenor Firmin: Haitian pioneer of anthropology," *American Anthropologist* 102, No. 3 (2000): 449–466.
Fluehr-Lobban, C., "Anténor Firmin and Haiti's contribution to anthropology," *Gradhiva: Revue d'Anthropologie et de histoire des Arts* 1 (2005): 95–108.
Fodor, Jerry, *Modularity of Mind* (Cambridge: MIT Press, 1983).
Foer, Joshua, "It's time for a conversation," *National Geographic* 277, No. 5 (May 2015): 1–14.
Foley, R., "Evolutionary perspectives on the origins of human social institutions," *Proceedings of the British Academy* 110 (2001): 171–195.
Forel, A., *The Social World of the Ants Compared with That of Man*, translated by C.K. Albert Ogden and Charles Boni (New York: G.P. Putnam's Sons, 1928).
Fortune, Reo, *Sorcerers of Dobu: The Social Anthropology of the Dobu Islanders of the Western Pacific* (New York: E.P. Dutton, 1932).

Fossum, John C., "Rent withholding and the improvement of substandard housing," *California Law Review* 53, No. 1 (March 1965): 304–336.
Foster, George M., "Peasant society and the image of limited good," *American Anthropologist, New Series* 67, No. 2 (1965): 293–315.
Fox, Robin, "Prejudice and the unfinished mind: a new look at an old failing," *Psychological Inquiry* 3, No. 2 (1992): 137–152.
Foucault, Michel, *Discipline and Punish* (New York: Vintage Books, 1991). Originally published in 1977.
Fowler, Henry, and Frank Fowler, *The King's English* (Oxford: Oxford University Press, 1906).
Fowler, Henry, *A Dictionary of Modern English Usage* (Oxford: Clarendon Press, 1926).
Fraenkel, Gottfried, and Donald L. Gunn, *The Orientation of Animals* (New York: Dover, 1961).
Fraga, M.F., E. Ballestar, M.F. Paz et al., "Epigenetic differences arise during the lifetime of monozygotic twins," *Proceedings of the National Academy of Sciences, USA* 102 (2005): 10604–10609.
Franklin, Elizabeth L., Elva J.H. Robinson, James A.R. Marshall, Ana B. Sendova-Franks, and Nigel R. Franks, "Do ants need to be old and experienced to teach?" *Journal of Experimental Biology* 215 (2012): 1287–1292. doi: 10.1242/jeb.064618.
Franks, N.R., J.P. Stuttard, C. Doran et al., "How ants use quorum sensing to estimate the average quality of a fluctuating resource," *Scientific Reports* 5 (2015): 11890.
Fredricks, D.N., and D.A. Relman, "Sequence-based identification of microbial pathogens: a reconsideration of Koch's postulates," *Clinical Microbiology Reviews* 9, No. 1 (January 1996): 18–33.
Freese, Barbara, *Coal: A Human History* (London: Penguin Books, 2004).
Freud, Sigmund, *The Complete Introductory Lectures on Psychoanalysis*, translated and edited by James Strachey (New York: W.W. Norton, 1966).
Fried, Morton H., *The Notion of Tribe* (Menlo Park: Cummings, 1975).
Von Frisch, Karl, *The Dancing Bees* (New York: Harcourt, Brace, Jovanovich, 1953).
Von Frisch, Karl, *The Dance Language and Orientation of Bees* (Cambridge: Harvard University Press, 1967).
Von Frisch, K., *Bees: Their Vision, Chemical Senses and Language*, revised edition (Ithaca: Cornel Press, 1971).
Fromkin, Victoria, and Robert Rodman, *An Introduction to Language* (New York: Holt, Reinhart and Winston, 1978).
Fuentes, Agustin, "Integrative anthropology and the human niche: toward a contemporary approach to human evolution," *American Anthropologist* 117, No. 2 (June 2015): 302–315.
Furst, Peter, *Flesh of the Gods: The Ritual Use of Hallucinogens* (Crows Nest, Australia: Allen & Unwin, 1972).
Furst, P., *Hallucinogens and Culture* (Novato, CA: Chandler & Sharp, 1977).
Gamble, G., *Timewalkers: The Prehistory of Global Colonization* (London: Penguin Books, 1993).

Gamo, N.T., and A.F. Amsten, "Molecular modulation of prefrontal cortex: rational development of treatments for psychiatric disorders," *Behavioral Neuroscience* 125, No. 3 (June 2011): 282–296.

Gardner, Howard, *Frames of Mind: The Theory of Multiple Intelligences* (Boston: Basic Books, 1983).

Gardner, Howard, and Katie Davis, *The App Generation* (New Haven: Yale University Press, 2013).

Garland, A., and J. Low, "Addition and subtraction in wild New Zealand robins," *Behavioral Processes* 109, Part B (November 2014): 103–110.

Gellner, Ernest, *Postmodernism, Reason and Religion* (London: Routledge, 1992).

Gellner, E., "Culture, constraint and community: semantic and coercive compensations for the genetic under-determination of *Homo sapiens sapiens*," in *The Human Revolution*, edited by Paul Mellars and Chris Stringer (Princeton: Princeton University Press, 1989): 314–328.

Gerringer, Steven, "The king and 'I'," *Practical Campbell* (Joseph Campbell Foundation, 2006). http://www.jcf.org/new/downloads/essays/Practical-Campbell_20060611_King-and-I.pdf, accessed January 3, 2017.

Geschwind, D.H., and P. Rakic, "Cortical evolution: judge the brain by its cover," *Neuron* 80 (2013): 633–647.

Geschwind, Norman, "The development of the brain and the evolution of language," in *Monograph Series on Language and Linguistics*, vol. 17, edited by C.I.J.M. Stewart (Washington: Georgetown University Press, 1964): 155–169.

Geschwind, N., "The organization of language and the brain," *Science* 170 (November 20, 1970): 940–944.

Ghafouri-Fard, S., M. Fardaei, M. Gholami, and M. Miryounesi, "A case report: autosomal recessive microcephaly caused by a novel mutation in MCPH1 gene," *Gene* 571, No. 1 (October 15, 2015): 149–150.

Gibbons, Ann, "Pleistocene population explosions," *Science* 262, No. 5130 (October 1993): 27–28.

Gibson, McGuire, "Violation of fallow and engineered disaster in Mesopotamian Civilization," in *Irrigation's Impact on Society*, edited by T.E. Downing and McG. Gibson (Tucson, AZ: The University of Arizona Press, 1974).

Gillooly, James, Chen Hou, and Michael Kaspari, "Eusocial insects as superorganisms: insights from metabolic theory," *Communicative and Integrative Biology* 3, No. 4 (July–August 2010): 360–362.

Gimbutas, Marija, *The Language of the Goddess* (San Francisco: HarperSanFrancisco, 1989).

Glaeser, Edward, *Triumph of the City: How Our Greatest Invention Makes Us Richer, Smarter, Greener, Healthier and Happier* (New York: Penguin Press, 2011).

de Gobineau, Joseph Arthur, *Essai sur l'Inegalite des Races Humaines*, vol. 4 (Paris: Belford, 1853–1855).

de Gobineau, Joseph Arthur, *Correspondance entre le comte de Gobineau et le comte de Prokesch-Osten (1854–1876)* (Paris: Plon, 1933).

Godzinska, Ewa Joanna, "Human and ant social behavior should be compared in a very careful way to draw valid parallels," *Behavioral and Brain Science* 39 (2016): 21–22.

Goldenweiser, Alexander, *Early Civilization* (New York: Knopf, 1922).

Goldin-Meadow, Susan, Carolyn Mylander, Jill de Villiers et al., "Gestural communication in deaf children: the effects and noneffects of parental input on early language development," *Monographs of the Society for Research in Child Development* 49, No. 3/4 (1984): 1–151.

Goldman, Alvin I., "A causal theory of knowing," *Journal of Philosophy* 64, No. 12 (June 22, 1967): 357–372.

Goldman, S.A., M. Nedergaard, and M.S. Windrem, "Modeling cognition and disease using human glial chimeric mice," *Glia* 63, No. 8 (August 2015): 1483–1493.

Gonzalez-Voyer, Alejandro, Svante Winberg, and Niclas Kolm, "Social fishes and single mothers: brain evolution in African cichlids," *Proceedings of the Royal Society of London B* 276 (2009): 161–167.

Goodnight, Charles J., "On the effectiveness of multilevel selection," *Behavioral and Brain Science* 39 (2016): 22–23.

Goody, Jack, *The Interface Between the Written and the Oral* (Cambridge: Cambridge University Press, 1987).

Goody, N.E., "Introduction: some implications of a social origin of intelligence," in *Social Intelligence and Interaction* (Cambridge: Cambridge University Press, 1995): 1–36.

Goos, Lisa M., and Gillian Ragesdale, "Genomic imprinting and human psychology: cognition, behavior and pathology," in *Genomic Imprinting*, edited by Jon F. Wilkins, Volume 626 of the series *Advances in Experimental Medicine and Biology* (New York: Springer, 2008): 71–88.

Gould, James L., "Why do honeybees have dialects?" *Behavioral Ecology and Sociology* 10, No. 1 (1982): 53–56.

Gould, S.J., *Ontogeny and Phylogeny* (Cambridge: Belknap Press, 1977).

Gould, Steven J., "Bligh's bounty; a captain who followed the rules meticulously found an animal that broke them brilliantly; the echidna," *Natural History* 94 (September 1985): 2.

Gould, Steven J., *Time's Arrow, Time's Cycle: Myth and Metaphor in the Discovery of Geological Time* (Cambridge: Harvard University Press reprint, 1988).

Gould, Steven J., *Full House: The Spread of Excellence from Plato to Darwin* (New York: Harmony Books, 1996a).

Gould, Steven J., *The Mismeasure of Man* (New York: W.W. Norton, 1996b). Originally published in 1981.

Govero, J., P. Esakky, S.M. Schaeaffer et al., "Zika virus infection damages the testes in mice," *Nature* 540 (October 31, 2016): 438–442. doi: 10.1038/nature20556.

Gowdy, John, and Lisi Krall, "Economic origins of ultrasociality," *Behavior and Brain Science* 39 (2016): 1–60.

Gowlett, John, Clive Gamble, and Robin Dunbar, "Human evolution and the archaeology of the social brain," *Current Anthropology* 53, No. 6 (December 2012): 693–722.

Gowlett, John A.J., and Richard W. Wrangham, "Earliest fire in Africa: towards the convergence of archaeological evidence and the cooking hypothesis," *Azania: Archaeological Research in Africa* 48, No. 1 (2013): 5–30.

Grasse, Pierre-Paul, *Parasites et parasitisme* (Paris: Armand Collin, 1935).

Graziano, M.S.A. *Consciousness and the Social Brain* (Oxford: Oxford University Press, 2013).

Greaves, Mel, *Cancer: The Evolutionary Legacy* (Oxford: Oxford University Press, 2000).

Green, A.J., J.R. Yates, A.M. Taylor et al., " Severe microcephaly with normal intellectual development: the Nijmegen Breakage syndrome," *Archives of Disease in Childhood* 73, No. 5 (November 1995): 431–434.

Greenbaum, Susan, Cheryl Rodriguez, and Beverly G. Ward, "Displacement and deconcentration in Tampa," *Anthropology News*, December 2008: 10–14.

Greenfield, Susan, *A Day in the Life of the Brain: The Neuroscience of Consciousness from Dawn till Dusk* (New York: Allen Lane, 2016).

Greer, Germaine, *The Female Eunuch* (London: Granada Publishing, 1970).

Gregory, Steven, and Roger Sanjek, *Race* (New Brunswick: Rutgers University Press, 1994).

Griffin, Donald R., *The Question of Animal Awareness: The Evolutionary Continuity of Mental Experience* (New York: Rockefeller University Press, 1976).

Griffin, Donald R., *Animal Minds: Beyond Cognition to Consciousness* (Chicago: Chicago University Press, 1992, revised 2001).

Grimaldi, G., and M. Manto, "Topography of cerebellar deficits in humans," *Cerebellum* 11, No. 2 (June 2012): 336–351. doi: 10.1007/s12311–011–0247–4.

Grinsted, L., I. Agnarsson, and T. Bide, "Social behavior and brood adoption in mixed-species colonies of two theridiid spiders," *Naturwissenschaften* 99 (2012): 1021–1030.

Groh, C., and W. Rossler, " Comparison of microglomerular structures in the mushroom body calyx of neopteran insects," *Arthropod Structure & Development* 40 (2011): 358–367.

Groh, Claudia, Lu Zhiyuan, Ian A. Meinertzhagen, and Wolfgang Rossler, "Age-related plasticity in the synaptic ultrastructure of neurons in the mushroom body calyx of the adult honeybee *Apis mellifera*," *Journal of Comparative Neurology Research in Systems Neuroscience* 520 (2012): 3509–3527.

Grosberg, R.K., and R.R. Strathmann, "The evolution of multicellularity: a minor major transition?" *Annual Review of Ecology Evolution and Systematics* 38 (2007): 621–654.

Gumilev, Lev, *Ethnogenesis and the Biosphere* (Moscow: Progress, 1990).

Gurnell, A.M., "The hydrogeomorphological effects of beaver dam-building activity", *Progress in Physical Geography* 22, No. 2 (June 1998): 167–190.

Gu, W., F. Zhang, and J.R. Lupski, "Mechanisms for human genomic rearrangements," *Pathogenetics* 1, No. 4 (2008): 1–4.

Gurven, Michael, and Hillard Kaplan, "Longevity among hunter-gatherers: a cross-cultural examination," *Population and Development Review* 33, No. 2 (June 2007): 321–365.

Guzewicz, Tony D., *Tokyo's Homeless: A City in Denial* (Huntington, NY: Kroshka Books, 2000).

Haldane, John Burton Sanderson, *The Causes of Evolution* (Ithaca: Cornell University Press, 1966 edition). Originally published in 1932.

Hall, Brian Keith, Benedikt Hallgrímsson, and Monroe W. Strickberger, *Strickberger's Evolution: The Integration of Genes, Organisms and Populations*, 4th edition (Mississauga: Jones and Bartlett, 2008).

Hall, E.T., *The Silent Language* (Garden City: Doubleday, 1959).

Hall, E.T., *The Hidden Dimension* (Garden City: Anchor Books, 1969).

Hall, Margaret I., Jason M. Kamilar, and E. Christopher Kirk, "Eye shape and the nocturnal bottleneck of mammals," *Proceedings of the Royal Society B* 1 (October 24, 2012): 4962–4968.

Halpern, Marnie E., Onur Gunturkun, William D. Hopkins, and Lesley J. Rogers, "Lateralization of the vertebrate brain: taking the side of model systems," *Journal of Neuroscience* 25, No. 45 (November 9, 2005): 10351–10357.

Harcourt, D.G., and E.J. Leroux, "Population regulation in insects and man," *American Scientist* 55 (1967): 400–415.

Harnack, Adolf, *What Is Christianity?* 2nd edition (New York: G.P. Putnam and Sons, 1902).

Harner, Michael, *Hallucinogens and Shamanism* (Oxford: Oxford University Press, 1973).

Harrell, Ashley, "A numbers game: S.F.'s flawed system for counting the homeless," *SF Weekly* (February 10, 2009): 10.

Harris, Marvin, *The Rise of Anthropological Theory: A History of Theories of Culture* (New York: Thomas Y. Crowell, 1968).

Harrison, K. David, *When Languages Die* (Oxford: Oxford University Press, 2007).

Hart, Donna, Robert Wald Sussman, and Robert W. Sussman, *Man the Hunted: Primates, Predators, and Human Evolution* (New York: Basic Books, 2005).

Hartline, Daniel K., "What is myelin?" *Neuron Glia Biology* 4 (2008): 153–163.

Hartmann, Chester, and A. Averbach, *Yerba Buena: Land Grab and Community Resistance in San Francisco* (San Francisco: Glide, 1974).

Hartsuiker, Dolf, *Sadhus: India's Mystic Holy Men* (Rochester, VT: Inner Tradition, 1993).

Harvard Medical School, "Blue light has a dark side," *Harvard Health Letter* 2012, updated September 2015. http://www.health.harvard.edu/staying-healthy/blue-light-has-a-dark-side, accessed January 3, 2017.

Harvey, David, *Rebel Cities: From Right to the City to the Urban Revolution* (London: Verso, 2011).

Harvey, Fiona, "How bricks and mortar can help save the planet," *Financial Times*, January 16, 2006.

Harvey, Paul H., and Peter M. Bennett, "Brain size, energetics, ecology and life history patterns," *Nature* 306 (November 24, 1983): 314–315.

Haslam, Michael, Adriana Hernandez-Aguilar, Victoria Ling et al., "Primate archaeology," *Nature* 460 (July 16, 2009): 339–344.

Hatano, Giyoo, and Keiko Takahashi, "Is shared intentionality widespread among and unique to humans?" *Behavioral and Brain Sciences* 28, No. 5 (October 2005): 703.

Haug, Herbert, "Brain sizes, surfaces, and neuronal sizes of the cortex cerebri: a stereological investigation of man and his variability and a comparison with some

mammals (primates, whales, marsupials, insectivores, and one elephant)," *The American Journal of Anatomy* 180 (1987): 126–142.
Hauser, Marc D., Noam Chomsky, and W. Tecumseh Fitch, "The faculty of language: what is it, who has it, and how did it evolve?" *Science* 298, No. 5598 (November 22, 2002): 1569–1579.
Hawking, Steven, and Leonard Mlodinow, *The Grand Design* (New York: Bantam Books, 2010).
Healy, S.D., and J.R. Krebs, "Food storing and the hippocampus in corvids: amount and size correlated," *Proceedings of the Royal Society of London B* 248 (1992): 241–245.
Hegel, Georg Wilhelm Frederick, *The Philosophy of History* (New York: Dover 1956). Originally published in 1830–1831.
Heinze, Jurgen, and Walter Bartosz, "Moribund ants leave their nests to die in social isolation," *Current Biology* 20 (February 9, 2010): 249–252.
Hennah, William, and David Porteous, "The DISC1 pathway modulates expression of neurodevelopmental, synaptogenic and sensory perception genes," *PLOS One* 4, No. 3 (March 20, 2009).
Henry, Amanda G., Alison S. Brooks, and Dolores R. Piperno, "Plant foods and the dietary ecology of Neanderthals and early modern humans," *Journal of Human Evolution* 69 (April 2014): 44–54.
Henry, Jules, *Culture Against Man* (New York: Vintage Books, 1963).
Henson, Kenneth T., "Teaching methods: history and status," *Theory Into Practice* 19, No. 1, "Teaching methods: designs for learning" (Winter 1980): 2–5.
Herculano-Houzel, S., and R. Lent, "Isotropic fractionator: a simple, rapid method for the quantification of total cell and neuron numbers in the brain," *Journal of Neuroscience* 25, No. 10 (2005): 2518–2521.
Herculano-Houzel, S., C. Collins, P. Wong, and J. Kaas, "Cellular scaling rules for primate brains," *Proceedings of the National Academy of Sciences, USA* 104, No. 9 (2007): 3562–3567.
Herculano-Houzel, Suzana, Bruno Motal, and Pokerto Lent, "Cellular scaling rules for rodent brains," *Proceedings of the National Academy of Sciences, USA* 103, No. 32 (August 8, 2006): 12138–12143.
Herculano-Houzel, Suzana, "The human brain in numbers: a linearly scaled-up primate brain," *Frontiers in Human Neuroscience* 3, No. 31 (November 9, 2009): 1–11.
Herculano-Houzel, Suzana, Bruno Mota, Peivan Wong, and Jon H. Kaas, "Connectivity-driven white matter scaling and folding in primate cerebral cortex," *Proceedings of the National Academy of Sciences, USA* 107, No. 44 (November 2, 2010): 19008–19013.
Herculano-Housel, Suzana, "The remarkable, yet not extraordinary, human brain as a scaled-up primate brain and its associated cost," *Proceedings of the National Academy of Sciences, USA* 109, Suppl. 1 (June 26, 2012): 10661–10668.
Herculano-Houzel, Suzana, "Decreasing sleep requirement with increasing numbers of neurons as a driver for bigger brains and bodies in mammalian evolution." *Proceedings, Royal Society, B.* 282 (2015): 1853. doi.org/10.1098/rspb.2015.1853.

Herculano-Houzel, S., K. Catania, P.R. Manger, and J.H. Kaas, "Mammalian brains are made of these: a dataset of the numbers and densities of neuronal and non-neuronal cells in the brain of Glires, Primates, Scandentia, Eulipotyphlans, Afrotherians and Artiodactyls, and their relationship with body mass," *Brain, Behavior and Evolution* 86, No. 3–4 (2015), 1–19.

Herculano-Houzel, Suzana, Paul R. Manger, and Jon H. Kaas, "Brain scaling in mammalian evolution as a consequence of concerted and mosaic changes in numbers of neurons and average neuronal cell size," *Frontiers in Neuroanatomy* 77 (August 8, 2014): 1–28.

Herlihy, David, *Women, Family and Society in Medieval Europe: Historical Essays, 1978–1991* (Providence, RI: Berghahn Books, 1995).

Herschensohn, Julia, "Review of the language instinct," *The Modern Language Journal* 91, No. 3 (Autumn 2007): 486–487.

Herskovits, Melville, *Cultural Anthropology: An Abridged Edition of Man and His Works* (New York: Knopf, 1965).

Herrera, Ana M., Simone G. Shuster, Claire L. Perriton, and Martin J. Cohn, "Developmental basis of phallus reduction during bird evolution," *Current Biology* 23, No. 12 (June 17, 2013): 1065–1074.

Hewes, G.W., *Language Origins: A Bibliography*, 2nd edition, vols. 1 and 2 (The Hague: Mouton 1975).

Heydenreuter, Reinhard, "Konig Ludwig I und der Fall Kaspar Hauser", in *Staat und Verwaltung in Bayern, Festschrift fur Wilhelm Volkert zum 75. Geburtstag (Schriftenreihe zur bayerischen Landesgeschichte),* edited by Konrad Ackermann and Alois Schmid (Munich: Beck, 2003): 465–476.

Higbie, Frank Tobias, *Indispensable Outcasts: Hobo Workers and Community in the American Midwest 1880–1930* (Urbana: University of Illinois Press, 2003).

Hill, K., M. Barton, and M.A. Hurtado, "The emergence of human uniqueness: characters underlying behavioral modernity," *Evolutionary Anthropology* 18 (2009): 187–200.

Hinshaw, Stephen P., Richard M. Scheffler, Brent Fulton et al., "International variation in treatment procedures for ADHD: social context and recent trends," *Psychiatric Services* 62, No. 5 (May 2011): 459–464.

Hochner, B., T. Shomrat, and G. Fiorito, "The octopus: a model for a comparative analysis of the evolution of learning and memory mechanisms," *Biological Bulletin* 210, No. 3 (June 2006): 308–317.

Hodder, Ian, "Simple orrelations between material culture and society, a review," in *The Spatial Organization of Culture*, edited by Ian Hodder (London: Duckworth, 1978): 3–24.

Hodder, Ian, *The Leopard's Tale: Revealing the Mysteries of Catalhoyuk* (London: Thames and Hudson, 2011).

Hodgson, Geoffrey M., "What are institutions?" *Journal of Economic Issues* XL, No. 1 (2006): 1–25.

Hoebel, E. Adamson, *The Law of Primitive Man* (Cambridge: Harvard University Press, 1964).

Hofman, Michel, "Encephalization in hominids: evidence for the model of punctuationalism," *Brain, Behaviour and Evolution* 22, No. 2–3 (February 1983): 102–117.
Hofman, Michel A., "Evolution of the human brain: when bigger is better," *Frontiers in Neuroanatomy* 8, article 15 (March 2014): 1–12.
Hofman, Michel A., and Dean Falk, *Evolution of the Primate Brain: From Neuron to Behavior* (Amsterdam: Elsevier, March 2, 2012): 425.
Hoffmann, J.A., F.C. Kafatos, C.A. Janeway, Jr., and R.A.B. Ezekowitz, "Phylogenetic perspectives in innate immunity," *Science* 284 (1999): 1313–1318.
Holldobler, Bert, and E.O. Wilson, (1977) "Colony-specific territorial pheromone in the African weaver ant *Oecophylla longinoda* (Latreille)," *Proceedings of the National Academy of Sciences, USA* 74, No. 5 (1999): 2072–2075.
Holldobler, Bert, and E.O. Wilson, *The Ants* (Cambridge: Belknap Press, 1990).
Hölldobler, Bert, and E.O. Wilson, *The Superorganism: The Beauty, Elegance, and Strangeness of Insect Societies* (New York: W.W. Norton, 2008).
Holloway, R.L., "On the meaning of brain size," a review of H.J. Jerison's 1973 *Evolution of the Brain and Intelligence, Science* 184 (1974): 677–679.
Holloway, R.L., *The Role of Human Social Behavior in the Evolution of the Brain* (New York: American Museum of Natural History, 1975).
Holloway, R.L., "Brain size, allometry, and reorganization: toward a synthesis," in *Development and Evolution of Brain Size: Behavioral Implications*, edited by M.E. Hahn, G. Jensen, and B.C. Dudek (New York: Academic Press, 1979): 59–88.
Holloway, R.L., and David G. Post, "The relativity of relative brain measures and hominid mosaic evolution," in *Primate Brain Evolution: Methods and Concepts*, edited by Este Armstrong and Dean Falk (New York: Plenum, 1982): 57–76.
Holloway, R.L., "Another primate brain fiction: brain (cortex) weight and homogeneity," *Behavioral and Brain Sciences* 16, No. 4 (1993): 107–108.
Holloway, R.L., "Evolution of the human brain," in *Handbook of Human Symbolic Evolution*, edited by A. Lock and C. Peters (New York: Oxford University Press, 1996): 74–116.
Holloway, R.L., D.C. Broadfield, and M.S. Yuan, *The Human Fossil Record, Volume Three: Brain Endocasts, The Paleoneurological Evidence* (New York: Wiley-Liss, 2004).
Holloway, R.L., "Neural parameters, hunting, and the evolution of the human brain," in *Advances in Primatology, Volume 1: The Primate Brain,* edited by C.R. Noback and W. Montagna (New York: Appleton-Century-Crofts, 1970): 299–310.
Holmberg, Allan R., *Nomads of the Long Bow* (Garden City: The Natural History Press, 1969). Originally published in 1950.
Holt, Lori L., and Andrew J. Lotto, "Speech perception as categorization," *Attention, Perception, & Psychophysics* 72, No. 5 (July 2010): 1218–1227.
Hoogland, J.L., *The Black-tailed Prairie Dog: Social Life of a Burrowing Mammal* (Chicago: The University of Chicago Press, 1995).
Horgan, Terrance, "Naturalism and intentionality," *Philosophical Studies* 76 (1994): 301–326.
Horn, Susanne, Walter Durka, Ronny Wolf et al., "Mitochondrial genomes reveal slow rates of molecular evolution and the timing of speciation in beavers (Castor),

one of the largest rodent species," *PLoS ONE* 6, No. 1 (January 28, 2011). doi: 10.1371/journal.pone.0014622.

Horner, Victoria, Kristen E. Bonnie, and Frans B.M. de Waal, "Identifying the motivations of chimpanzees: culture and cognition," *Behavioral and Brain Sciences* 28, No. 5 (2005): 704–705.

Hou, Chen, Michael Kaspari, Hannah B. Vander Zanden, and James F. Gillooly, "Energetic basis of colonial living in social insects," *Proceedings of the National Academy of Sciences, USA* 107, No. 8 (February 23, 2010): 3634–3638.

Houdek, Petr, Julie Novakova, and Dan Stastny, "Ultrasociality: when institutions make a difference," *Behavioral and Brain Sciences* 39 (2016): 24–26.

Hrdy, Sarah, "Male–male competition and infanticide among the langurs (*Presbytes entellus*) of Abu, Rajasthan," *Folio Primatologica* 22 (1974): 19–58.

Hrdy, Sarah, *The Woman That Never Evolved* (Cambridge: Harvard University Press, 1981).

Hrdy, Sarah, *Mother Nature: A History of Mothers, Infants and Natural Selection* (New York: Pantheon, 1999).

Hrdy, Sarah, *Mothers and Others: The Evolutionary Origins of Mutual Understanding* (Cambridge: Harvard University Press, 2009).

Hsu, Francis L.K., "Rethinking the concept, 'Primitive,'" *Current Anthropology* 5, No. 3 (June 1964): 169–178.

Hublin, Jean-Jacques, Abdelouahed Ben-Ncer, Shara E. Bailey et al., "New fossils from Jebel Irhoud, Morocco and the pan-African origin of *Homo sapiens,*" *Nature* 546 (June 8, 2017): 289–292.

Huett, Alan, Aylwin Ng, Zhifang Cao et al., "A novel hybrid yeast-human network analysis reveals an essential role for FNBP1L in antibacterial autophagy," *Journal of Immunology* 182, No. 8 (April 15, 2009): 4917–4930.

Huffard, C.L., F. Boneka, and R.J. Full, "Underwater bipedal locomotion by octopuses in disguise," *Science* 307 (2005): 1927.

Huizinga, J., *The Waning of the Middle Ages* (New York: Doubleday, 1954). Originally published in 1924.

Hulcr, Jiri, and Anthony Cognato, "Repeated evolution of crop theft in fungus-farming ambrosia beetles," *Evolution* 64, No. 11 (November 2010): 3205–3212.

Hulcr, Jiri, and L.L. Stelinsky, "The ambrosia symbiosis: from evolutionary ecology to practical management," *Annual Review of Entomology* 62 (2016): 285–303. doi: 10.1146/annurev-ento-031616–035105.

Hull, David L., "Nurturing a view of human nature," review of Steven Pinker's *The Blank Slate: The Modern Denial of Human Nature, Nature* 419 (September 19, 2002): 251–252.

Humphrey, N., "Vision in a monkey without striate cortex: a case study," *Perception* 3 (1974): 241–255.

Humphrey, N.K., "The social function of intellect," in *Growing Points in Ethnology*, edited by P.P.G. Bateson and R.A. Hinde (Cambridge: Cambridge University Press, 1976): 303–317.

Humphrey, N., *Seeing Red: A Study in Consciousness* (Cambridge: Belknap Press, 2006).

Huizinga, Johan, *Homo Ludens: A Study of the Play Element in Culture* (Boston: Beacon Books, 1938).
Hurt, R. Douglas, *Indian Agriculture in America: Prehistory to the Present* (Lawrence, KS: University Press of Kansas, 1987).
Husseman, Jason S., Dennis L. Murray, Gary Power, Curt Mack, C.R. Wenger, and Howard Quigley, "Assessing differential prey selection patterns between two sympatric large carnivores," *Oikos* 101, No. 3 (June 2003): 591–601.
Hutchinson, E.P., *The Population Debate* (Boston: Houghton Mifflin, 1967).
Huth, Alexander G., Wendy A. de Heer, Thomas L. Griffiths, Frederic E. Theunissen, and Jack L. Gallant, "Natural speech reveals the semantic maps that tile human cerebral cortex," *Nature* 532 (April 28, 2016): 453–460.
Hymes, D.H., *Ethnography, Linguistics, Narrative Inequality [Electronic Resource]: Toward an Understanding of Voice* (London: Taylor and Francis, 1996).
Ingham, Richard, "Input and learnability: direct-object omissibility in English," *Language Acquisition* 3, No. 2 (1993/1994): 95–120.
Isaac, G.L., "Food sharing and human evolution: archaeological evidence from the Plio-Pleistocene of East Africa," *Journal of Anthropological Research* 34 (1978): 311–325.
Isaac, G.L., "Evolutionary hypotheses," *Behavioral and Brain Sciences* 2 (1979): 388.
Isler, K., and C.P. van Schaik, "Metabolic costs of brain size evolution," *Biology Letters* 2 (2006): 557–560.
Jackson, Duncan E., Stephen J. Martin, Mike Holcombe, and Francis L.W. Ratnieks, "Longevity and detection of persistent foraging trails in Pharaoh's ants, *Monomorium pharaonis* (L.)," *Animal Behavior* 71, No. 2 (February 2006): 351–359.
Jackson, A.P., H. Eastwood, S.M. Bell et al., "Identification of microcephalin, a protein implicated in determining the size of the human brain," *American Journal of Human Genetics* 71, No. 1 (July 2002): 136–142.
Jacob, T., E. Indriati, R.P. Soejono et al., "Pygmoid Australomelanesian *Homo sapiens* skeletal remains from Liang Bua, Flores: population affinities and pathological abnormalities." *Proceedings of the National Academy of Sciences, USA* 103, No. 36 (September 5, 2006): 13421–13426.
Jacobs, Jane, *The Economy of Cities* (New York: Random House, 1969).
Jacobson, Ken, "ADHD in cross-cultural perspective: some empirical results," *American Anthropologist* 104, No. 1 (March 2002): 8–11.
Jacobson, Thorkild, "Primitive democracy in ancient Mesopotamia," *Journal of Near Eastern Studies* 2, No. 3 (1943): 159–172.
Jargowsky, Paul, *Architecture of Segregation: Civil Unrest, the Concentration of Poverty, and Public Policy, Report: Race and Inequality* (New York: The Century Foundation, 2015).
Jaspers, Karl, *The Origin and Goal of History* (New Haven: Yale University Press, 1953).
Jeffrelot, C., *Dr. Ambedkar and Untouchability: Fighting the Caste System* (New York: Columbia University Press, 2005).
Jellinek, Michael S., "Psychiatric drugs and children," *Letters, Science* 300, No. 5621 (May 9, 2003): 901.

Jencks, Charles, *Architecture 2000 and Beyond* (New York: Wiley, 2000).
Jencks, Charles, *Critical Modernism* (New York: Wiley, 2007).
Jencks, Christopher, *The Homeless* (Cambridge: Harvard University Press, 1994).
Jerison, H.J., *Evolution of Brain and Intelligence* (New York: Academic Press, 1973).
Jerison, H.J., "Animal intelligence as encephalization," *Philosophical Transactions of the Royal Society B* 308 (1985): 21–35.
Johnson, Brian R., and Elisabeth Frost, "Individual-level patterns of division of labor in honeybees highlight flexibility in colony-level developmental mechanisms," *Behavioral Ecology and Sociobiology* 66 (2012): 923–930. doi: 10. 1007/s00265–0 12–1341–4.
Johnson, David Cay, "The hidden cost of the world's 'Ghost Apartments'," *Newsweek*, April 14, 2015.
Johnson, Eris A., Ricardo D. Salvatore, and Pieter Spierenburg, "Murder and mass murder in pre-modern Latin America: from pre-Colonial Aztec sacrifices to the end of Colonial rule, an introductory comparison with European societies," *Historical Social Research* 37, No. 3 (2012): 233–253.
Johnson, Linda A., "Doctors worry as drug-resistant staph infects more healthy people," *The Marin Independent Journal* (September 30, 2004): A7.
Jolij, Jacob, Victor A.F. Lamme, and Michael I. Posner, "Repression of unconscious information by conscious processing: evidence from affective blindsight induced by transcranial magnetic stimulation," *Proceedings of the National Academy of Sciences, USA* 102, No. 30 (July 26, 2005): 10747–10751.
Jolly, Clifford J., "The seed-eaters: a new model of hominid differentiation based on a baboon analogy," *Man*, New Series 5, No. 1 (March 1970): 5–26.
Jones, F.C., M.G. Grabherr, Y.F. Chan et al., Broad institute genome sequencing platform and whole genome assembly team, "The genomic basis of adaptive evolution in threespine sticklebacks," *Nature* 484 (April 5, 2012): 55–61.
Jones, K.E., and A.M. MacLarnon, "Affording affordably larger brains: testing hypotheses of mammalian brain evolution in bats," *American Naturalist* 164 (2004): 20–31.
Jones, Kenneth Lyons, *Smith's Recognizable Patterns of Human Malformation*, 4th edition (Philadelphia: Saunders Elsevier, 1988).
Jones, Thomas C., and Patricia G. Parker, "Delayed juvenile dispersal benefits both mother and offspring in the cooperative spider *Anelosimus studiosus* (Araneae: Theridiidae)," *Behavioral Ecology* 13 (2002): 142–148.
Jonsson, E.G., L. Flyckt, E. Burgert et al., "Meta-analysis of the dopamine D3 receptor gene (DRD3) Ser9Gly variant and schizophrenia," *Psychiatric Genetics* 14, No. 9 (2004): 9–12.
Joordens, Josephine C.A., Francesco d'Errico, Frank P. Wesselingh et al., *"Homo erectus* at Trinil on Java used shells for tool production and art," *Nature* 518 (February 12, 2015): 228–231 (S1–13).
Joseph, Robert, *Neuropsychiatry, Neuropsychology, Clinical Neuroscience*, 2nd edition (Baltimore: Williams and Wilkins, 1996).
Kaas, J.H. "The evolution of brains from early mammals to humans," *Wiley Interdiscplinary Reviews: Cognitive Science* 4 (2013): 33–45.
Kagan, Jerome, "Attention and psychological change in the young child," *Science* 170 (November 20, 1970): 826–832.

Kamin, L.J., "The pioneers of IQ testing," in *The Bell Curve Debate: History, Documents, Opinions*, edited by Russell Jacoby and Naomi Glauberman (New York: Times Books, 1995a): 476–508.

Kamin, L.J., "Behind the curve," *Scientific American* 272 (February 1995b): 99–103.

Kamin, Leon J., and A.S. Goldberger, "Twin studies in behavioral research: a skeptical view," *Theoretical Population Biology* 61 (2002): 83–95.

Kang, J., I. Jiang, S.A. Goldman, and M. Nedergaard, "Astrocytes-mediated potentiation of inhibitory synaptic transmission," *Nature Neuroscience* 1 (1998): 683–692.

Kant, Immanuel, *Prolegomena to Any Future Metaphysics That Will Be Able to Present Itself as a Science,* translation from the German based on the original editions with an introduction and notes by Peter G. Lucas (Manchester: Manchester University Press, 1953). Originally published in 1783.

Kaplan, Robert, *The Nothing That Is: A Natural History of Zero* (Oxford: Oxford University Press, 2000).

Kappelman, John, "The evolution of body mass and relative brain size in fossil hominids," *Journal of Human Evolution* 30 (1996): 243–276.

Karmiloff-Smith, Annette, *Beyond Modularity: A Developmental Perspective on Cognitive Science* (Cambridge: MIT Press, 1996).

Kasting, J.F., "Runaway and moist greenhouse atmospheres and the evolution of Earth and Venus," in *Planetary Sciences, American and Soviet Research Proceedings from the US/USSR Workshop on Planetary Sciences, Commission on Engineering and Technical Systems (CETS)* (1991): 234–245.

Katz, D.L., and S. Meller, "Can we say what diet is best for health?" *Annual Review of Public Health* 35 (March 2014): 83–103.

Kawai, M., "Newly-acquired pre-cultural behavior of the natural troop of Japanese monkeys on Koshima Islet," *Primates* 6 (1965): 1–30.

Keita, S.O.Y., "Further studies of crania from ancient Northern Africa: an analysis of crania from First Dynasty Egyptian tombs, using multiple discriminant functions," *American Journal of Physical Anthropology* 87, No. 3 (1992): 245–254.

Keith, Nathaniel Schneider, *Politics and the Housing Crisis since 1930* (New York: Universe Books, 1973).

Kelava, I., E. Lewitus, and W.B. Huttner, "The secondary loss of gyrencephaly as an example of evolutionary phenotypical reversal," *Frontiers in Neuroanatomy* 7, No. 16 (June 26, 2013). http://dx.doi.org/10.3389/fnana.2013.00016, accessed January 4, 2017.

Keller, L., "Genetics of behavior," *Philosophical Transactions of the Royal Society B* 364 (2009): 3209–3216.

Keller, R., "The sociology of knowledge approach to discourse (SKAD)," *Human Studies* 34, No. 1 (2011): 43–65.

Kennair, Leif Edward Ottesen, "Human nature and the limits of blank slatism," *Human Nature Review* 2 (2002): 483–491.

Kennedy, Paul, *Preparing for the 21st Century* (New York: Random House, 1993).

Kenyatta, Jomo, *Facing Mt. Kenya* (New York: Vintage Books, 1938).

Kerth, G., N. Perony, and F. Schweitzer, "Bats are able to maintain long-term social relationships despite the high fission-fusion dynamics of their groups," *Proceedings of the Royal Society of London B* 278 (2011): 2761–2767. doi: 10.1098/rpsb.2010.2718.

Kerth, G., "Causes and consequences of sociality in bats," *Bioscience* 58 (2008): 737–755.

Keynes, John Maynard, *A Treatise on Probability* (Cambridge: Cambridge University Press, 1921).

Kimler, William, "Case studies, controversy and the 'Fieldworker's Regress'," *Annals of Science* 69, No. 1 (2012). doi: 10.1080/00033790.2010.510939.

Klein, Richard G., *The Human Career, Human Biological and Cultural Origins* (Chicago: University of Chicago Press, 1989).

Klein, Richard G., and Blake Edgar, *The Dawn of Human Culture* (San Francisco: John Wiley and Sons, 2002).

Klein, Jan, Naoyuki Takahata, and Francisco Ayala, "MHC polymorphism and human origins," *Scientific American* 269, No. 6 (December 1993): 78–83.

Knight, H. (2014). Why S.F.'s plan to end homelessness failed. San Francisco Chronicle, 29 June.

Koehler, O., "The ability of birds to 'count,'" *Bulletin of Animal Behavior* 9 (1950): 41–45.

Kotkin, Joel, *The City: A Global History* (New York: Modern Library, 2005).

Kramer, Samuel Noah, *Sumerian Mythology* (New York: Harper, 1961).

Krantz, G.S., "Pithecanthropine brain size and its cultural consequences," *Man* 11, No. 103 (1961): 85–87.

Krantz, G.S., "Brain size and hunting ability in earliest man," *Current Anthropology* 9, No. 5 (1968): 450–451.

Kraus, Cornelia, Cornelia van Waveren, and Franziska Huebner, "Distractible dogs, constant cats? A test of the distraction hypothesis in two domestic species," *Animal Behaviour* 93 (2014): 173–181.

Kripke, Saul, *Naming and Necessity* (Cambridge: Harvard University Press, 1980).

Kroeber, A.L., "The superorganic," *American Anthropologist* 19, No. 2 (1917): 116–123.

Krotoski, Aleks, "Robin Dunbar: we can only ever have 150 friends at most," *Observer*, March 14, 2010.

Kroutil, L.A., D.L. Van Brunt, M.A. Herman-Stahl et al., "Non-medical use of prescription stimulants in the USA," *Drug and Alcohol Dependence* 84 (February 14, 2006): 135–143.

Kruska, D.C.T., "The effects of domestication on brain size," in *Evolution of Nervous Systems* vol. 3, edited by Jon Kaas (New York: Academic Press, 2007): 143–153.

Kubo, K., O. Arai, M. Omura et al., "Low dose effects of bisphenol A on sexual differentiation of the brain and behavior in rats," *Neuroscience Research* 45, No. 3 (2003): 345–356.

Kucharski, R., J. Maleszka, S. Foret, and R. Maleszka, "Nutritional control of reproductive status in honey bees via DNA methylation," *Science* 319, No. 5871 (2008): 1827–1830.

Kuczaj, Stan A., and Lauren E. Highfill, "Dolphin play: evidence for cooperation and culture?" *Behavioral and Brain Sciences* 28, No. 5 (2005): 705–706.

Kuhlmeier, Valerie A., and Susan A.J. Birch, "Steps toward categorizing motivation: abilities, limitations and conditional constraints," *Behavioral and Brain Sciences* 28, No. 5 (2005): 706–707.

Kuruppuarachchi, K.A.L.A., and L.T. Wijeratne, "ADD in developing countries," *The British Journal of Psychiatry* 185, No. 5 (2004): 439. doi: 10.1192/bjp.185.5.439.

Laing, R.D., *The Divided Self* (Harmondsworth: Penguin Books, 1960). Originally published in 1959.

Laing, R.D., H. Phillipson, and A.R. Lee, *Interpersonal Perception: A Theory and a Method of Research* (New York: Perrenial Library, Harper and Row, 1966).

Laland, Kevin N., and Michael J. O'Brien, "Niche construction theory and archaeology," *Journal of Archaeological Method and Theory* 17 (2010): 303–322.

Lamberg-Karlovsky, C.C., and Jeremy A. Savloff, *The Rise and Fall of Civilizations: Modern Archaeological Approaches to Ancient Cultures* (Menlo Park: Cummings, 1974).

Lancaster, Carol, *Aid to Africa: So Much To Do, So Little Done* (Chicago: Chicago University Press, 1999).

Landes, David S., *The Wealth and Poverty of Nations: Why Some Nations Are So Rich and Some So Poor* (New York: W.W. Norton, 1999).

Langham, Ian, and David Oldroyd, *The Wider Domain of Evolutionary Thought* (London: Reidel, 1983).

Lashley, K.S., *Brain Mechanisms and Intelligence: A Quantitative Study of Injuries to the Brain* (Chicago: Chicago University Press, 1929).

Laskin, V.A., "Seasonal changes in the brain and skull sizes of small mammals," *Zhurnal obshcheĭ biologii* 50, No. 4 (1989): 470–480.

Laskowski, Kate L., and Jonathan N. Pruitt, "Evidence of social niche construction: persistent and repeated social interactions generate stronger personalities in a social spider," *Proceedings of the Royal Society B* 281 (2014): 20133166. http://dx.doi.org/10.1098/rspb.2013.3166, accessed January 4, 2017.

Lawlor, Anthony, *The Temple in the House: Finding the Sacred in Everyday Architecture* (New York: Tarcher Putnam, 1994).

Lawrence-Zuniga, Denise, and Deborah Pellow, "The self-consciousness of placemaking," *Anthropology News* (December 2008): 4–5.

Leakey, Louis S.B., *Adam's Ancestors: The Evolution of Man and His Culture* (New York: Harper Torchbooks, 1960). Originally published in 1934.

Lee, Dorothy, *Freedom and Culture* (Upper Saddle River, NJ: Prentice-Hall, 1959).

Lee, Richard B., "Power and property in twenty-first century foragers: a critical examination," in *Power and Equality: Encapsulation, Commercialization, Discrimination*, edited by T. Widlok and T. Wolde (Oxford: Berg, 2004): 16–31.

Lee, Sang-Hee, and Milford H. Wolpoff, "The pattern of evolution in Pleistocene human brain size," *Paleobiology* 29, No. 2 (2003): 186–196.

Lefebvre, L., P. Whitle, E. Lascaris, and A. Finkelstein, "Feeding innovations and forebrain size in birds," *Animal Behavior* 53 (1997): 549–560.

Leinberger, Christopher B., *The Option of Urbanism: Investing in a New American Dream* (Washington, DC: Island Press, 2007).

Lenneberg, Eric, *Biological Foundations of Language* (New York: Wiley, 1967).

Lenzer, Jeanne, "Truly independent research?" *British Medical Journal* 337 (September 13, 2008): 602–606.

Levine, Hal B., "Social insects, merely a 'fun house' mirror of human social evolution," *Behavioral and Brain Sciences* 39 (2016): 28.

Levinson, S.C., "Interactional biases in human thinking," in *Social Intelligence and Interaction*, edited by N.E. Goody (Cambridge: Cambridge University Press, 1995): 221–260.

Leviton, Alan, Lewis B. Holmes, Elizabeth N. Allred, and Juan Vargas, "Methodologic issues in epidemiologic studies of congenital microcephaly," *Early Human Development* 69, No. 1–2 (October 2002): 91–105.

Levy-Bruhl, L., *Primitive Mentality* (Boston: Beacon Press, 1923).

Lewis, Brian, *The Sargon Legend: A Study of the Akkadian Text and the Tale of the Hero Who Was Exposed at Birth* (Cambridge: American Schools of Oriental Research, 1980).

Lewis, Nathan, "Japan's greying population needs new solutions," *Forbes Magazine*, September 10, 2014.

Liberman, A.M., K.S. Harris, H.S. Hoffman, and B.C. Griffith, "The discrimination of speech sounds within and across phoneme boundaries," *Journal of Experimental Psychology* 54, No. 5 (1957): 358–368. doi: 10.1037/h0044417.

Liebenberg, Louis, "Persistence hunting by modern hunter-gatherers," *Current Anthropology* 47, No. 6 (December 2006): 1017–1026.

Lieberman, Paul, *The Biology and Evolution of Language* (Cambridge: Harvard University Press, 1984).

Lilly, John Cunningham, *Man and Dolphin* (Garden City: Doubleday, 1961).

Lin, Norman, and Charles D. Michener, "Evolution of sociality in insects," *The Quarterly Review of Biology* 47, No. 2 (1972): 132–159.

Lindauer, M., *Communication Among Social Bees*, revised edition (Cambridge: Harvard University Press, 1971).

Lindsay, Hugh, *Adoption in the Roman World* (Cambridge: Cambridge University Press, 2009).

Little, William, *Oxford Universal Dictionary*, revised and edited by C.T. Onions (Oxford: Oxford University Press, 1955).

Lloyd, G.E.R., *Magic, Reason and Experience: Studies in the Origin and Development of Greek Science* (Cambridge: Cambridge University Press, 1979).

Local Government Association (UK), "Demand for social housing fueled by recession," LGA Press Release, 2009.

Loehle, Craig, "Social barriers to pathogen transmission in wild animal populations," *Ecology* 76, No. 2 (March 1995): 326–335.

Longhurst, John Edward, *The Age of Torquemada* (Lawrence, KA: Doronado Press, 1964).

Lopez-Bao, José V., Alejandro Rodríguez, Miguel Delibes et al., "Revisiting food-based models of territoriality in solitary predators," *Journal of Animal Ecology* 83, No. 4 (April 2014): 934–942.

Louchart, Antoine, and Laurent Viriot, "From snout to beak: the loss of teeth in birds," *Trends in Ecology and Evolution* 26, No. 12 (December 2011): 663–673.

Lovejoy, C.O., "The origin of man," *Science* 211 (1981): 341–350.
Lovelock, James, *The Ages of Gaia* (New York: W.W. Norton, 1988).
Lowie, Robert, *History of Ethnological Theory* (New York: Holt, Rinehart and Winston, 1937).
Lowie, Robert, *Culture and Ethnology* (New York: Basic Books, 1966 edition). Originally published in 1917.
Lucentini, Jack, "Gene association studies typically wrong," *The Scientist*, December 20, 2004: 20.
Ludvig, Elliot A., "Why Pinker needs behaviorism: a critique of the Blank Slate," *Behavior and Philosophy* 31 (2003): 139–143.
Lumsden C.J., and E.O. Wilson, *Genes, Mind and Culture: The Coevolutionary Process* (Cambridge: Harvard University Press, 1981).
Lyn, Tan Ee, and Vicki Kwong, "HK seals off apartment building as virus spreads," *Reuters News Service*, March 31, 2003.
Lynch, P.A., and J. Roberts, *Native American Mythology A to Z*, (New York: Infobase, 2010).
Lyons, Derek E., Webb Phillips, and Laurie R. Santos, "Motivation is not enough: a commentary on Tomasello et al.,'s 'Understanding and sharing intentions: The origins of cultural cognition,'" *Behavioral and Brain Sciences* 28, No. 5 (2005): 708.
Macphail, E.M., *Brain and Intelligence in Vertebrates* (New York: Oxford University Press, 1982).
Macphail, E.M., "The comparative psychology of intelligence," *Behavioral and Brain Sciences* 10 (1987): 645–695.
Macphail, E.M., and Johan J. Bolhuis, "The evolution of intelligence: adaptive specializations vs general process," *Biological Reviews* 76 (2001): 341–364.
Machiavelli, Niccolo, *Florentine Histories* (Princeton: Princeton University Press, 1988), translated by Laura F. Banfield and Harvey Mansfield.
Mackintosh, Nicholas J., *IQ and Human Intelligence* (Oxford: Oxford University Press, 1998).
MacLean, Evan, Brian Hare, Charles Nunn et al., "The evolution of self-control," *Proceedings of the National Academy of Sciences, USA* 111, No. 20 (April 21, 2014): E2140–E2148.
Madden, Joah, "Sex, bowers and brains," *The Royal Society, Proceedings: Biological Sciences* 268, No. 1469 (April 22, 2001): 833–838.
Magalhaes, Sara, Arne Janssen, Marta Monserrat, and Maurice Sabelis, "Prey attack and predators defend: counterattacking prey trigger parental care in predators," *Proceedings of the Royal Society B* 272 (2005): 1929–1933.
Magnusson, Magnus, and Hermann Palsson, translators, *Laxdaela Saga* (New York: Penguin Books, 1969).
Maguire, E.A., D.G. Gadian, I.S. Johnsrude et al., "Navigation-related structural change in the hippocampi of taxi drivers," *Proceedings of the National Academy of Sciences, USA* 97 (2000): 4398–4403.
Makin, John H., "Housing and American Recessions," American Enterprise Institute Outlook Series, *Financial Services*, November 30, 2006.
Maloney, Thomas N., and Kim Korinek, editors, *Migration in the 21st Century* (London: Routledge, 2011).

Malinowski, B., *Magic, Science and Religion*, 1954 edition (New York: Anchor Books, 1916).

Malinowski, B., *Argonauts of the Western Pacific: An Account of Native Enterprise and Adventure in the Archipelagoes of Melanesian New Guinea* (London: Routledge and Kegan Paul, 1922).

Malinowski, B., *The Sexual Life of Savages* (London: Routledge and Kegan Paul, 1929).

Malinowski, B., "The language of magic and gardening," in *Coral Gardens and Their Magic*, vol. 2 (Bloomington: Indiana University Press, 1935).

Manger, P.R., "An examination of cetacean brain structure with a novel hypothesis correlating thermoregenesis in the evolution of a big brain," *Biological Reviews* 81 (2006): 293–338.

Marcus, Anthony, *Where Have All the Homeless Gone: The Making and Unmaking of a Crisis* (New York: Berghahn Books, 2006).

Marcus, J., "The peaks and valleys of ancient states: an extension of the dynamic model," in *Archaic States*, edited by G. Feinman and J. Marcus (Santa Fe: School of American Research, 1998): 59–94.

Margulis, Lynn, and Dorion Sagan, *What Is Life?* (New York: Simon and Schuster, 1995).

Margulis, Lynn, and Boris Mikhaylovich Kozo-Polyansky, *Symbiogenesis: A New Principle of Evolution* (Cambridge: Harvard University Press, 2010).

Margulis, Lynn, *Symbiosis in Cell Evolution: Life and Its Environment on the Early Earth* (San Francisco: W.H. Freeman, 1981).

Margulis, Lynn, *Symbiosis in Cell Evolution: Microbial Communities in the Archean and Proterozoic Eons* (San Francisco: W.H. Freeman, 1992).

Marino, L., "What can dolphins tell us about primate evolution?" *Evolutionary Anthropology* 5 (1996): 73–110.

Marino, Lori, Richard C. Connor, R. Ewan Fordyce et al., "Cetaceans have complex brains for complex cognition," *PLoS Biology*, May 15, 2007. doi: 10.1371/journal.pbio.0050139.

Markon, K.E., R.F. Krueger, T.J. Bouchard, and I.I. Gottesman, "Normal and abnormal personality traits: evidence for genetic and environmental relationships in the Minnesota study of twins reared apart," *Journal of Personality* 70, No. 5 (October 2002): 661–693.

Marriott, Alice, *Saynday's People: The Kiowa Indians and the Stories They Told* (Lincoln: University of Nebraska Press, 1963 edition). Originally published in 1947.

Martin, Gary E., Jessica Klusek, Bruno Estigarribia, and Joan E. Roberts, "Language characteristics of individuals with Down syndrome," *Topics in Language Disorders* 29, No. 2 (April 2009): 112–132.

Martin, Joanna, Miriam Cooper, Marian L. Hamshere et al., "Biological overlap of attention-deficit/hyperactivity disorder and autism spectrum disorder: evidence from copy number variants," *Journal of the American Academy of Child & Adolescent Psychiatry* 53, No. 7 (July 2014): 761–770.

Martin, R.D., "Human brain evolution in an ecological context," *52nd James Arthur Lecture on the Evolution of the Human Brain* (New York: American Museum of Natural History, 1983).

Martin, R.D., A.M. MacLarnon, J.L. Phillips, L. Dussubieux, P.R. Williams, and W.B. Dobyns, "Comment on 'The Brain of LB1, *Homo floresiensis*,'" *Science* 312, No. 5776 (May 19, 2006): 999.

Martin, R.D., M. Genoud, and C.K. Hemelrijk, "Problems of allometric saling analysis: examples from mammalian reproductive biology," *Journal of Experimental Biology* 208 (2005): 1731–1747.

Manson, J.H., and S. Perry, "Correlates of self-directed behaviour in wild white faced capuchins," *Ethology* 106, No. 4 (2000): 301–317.

Mathews, Gordon, *Ghetto at the Center of the World* (Chicago: University of Chicago Press, 2011).

Marzluff, John, *Gifts of the Crow* (New York: Atria, 2013).

McBrearty, Sally, and Allison S. Brooks, "The revolution that wasn't: a new interpretation of the origin of modern human behavior," *Journal of Human Evolution* 39 (2000): 453–563.

McCann, Donna, Angelina Barrett, Alison Cooper et al., "Food additives and hyperactive behavior in 3-year old and 8/9-year old children in the community: a randomized, double-blinded, placebo-controlled trial," *The Lancet* 370, No. 9598 (November 3, 2007): 1560–1567.

McClearn, G.E., Boo Johansson, Stig Berg et al., "Substantial genetic influence on cognitive abilities in twins 80 or more years old," *Science* 276 (1997): 1560–1563.

McDaniel, M.A., "Big brained people are smarter: a meta-analysis of the relationship between in vivo volume and intelligence," *Intelligence* 33 (2005): 337–346.

McClelland, D.C., *The Achieving Society* (New York: Van Nostrand, 1961).

McKaye, K.R., S.M. Louda, and J.R. Stauffer, Jr., "Bower size and male reproductive success in a cichlid fish lek," *American Naturalist* 135 (1990): 597–613.

McKenzie, Roderick D., *The Metropolitan Community* (New York: McGraw-Hill, 1933).

McLean, C.Y., P.L. Reno, A.A. Pollen et al., "Human-specific loss of regulatory DNA and the evolution of human-specific traits," *Nature* 471, No. 7337 (March 10, 2011): 216–219. doi: 10.1038/nature09774.

McNeill, William H., *Plagues and Peoples* (Garden City: Anchor Books, 1976).

Mead, Margaret, *Male and Female* (New York: William Morrow Books, 1949).

Mead, Margaret, *Continuities in Cultural Evolution* (Piscataway, NJ: Transaction, 1999). Originally published in 1965.

Mead, Margaret, *New Lives for Old: Cultural Transformation—Manus, 1928–1953* (New York: William Morrow, 1956).

Mead, Margaret, *Continuities in Cultural Evolution* (New Haven: Yale University Press, 1964).

Menzel, R., and M. Giurfa, "Cognitive architecture of a mini-brain: the honeybee," *Trends in Cognitive Sciences* 5 (2001): 62–71

Merbs, Charles F., "A new world of infectious disease," *American Journal of Physical Anthropology* 35 (1992): 3–42.

Mech, L.D., *The Wolf: The Ecology and Behavior of an Endangered Species* (New York: Natural History Press, 1970).

Merriam-Webster's Dictionary, 2017 edition (Merriam-Webster, Inc., 2017).

Meyer, Eric, and Sandra Duharcourt, "Epigenetic programming of developmental genome rearranges in ciliates," *Cell* 87 (October 4, 1996): 9–12.
Miller, E.K., "The prefrontal cortex: complex neural properties for complex behavior," *Neuron* 22 (1999): 15–17.
Miller, George A., "The magical number 7, plus or minus 2: some limits on our capacity for processing information," *Psychology Review* 63, No. 2 (1956): 81–97.
Miller, Geoffrey, *The Mating Mind: How Sexual Choice Shaped the Evolution of Human Nature* (New York: Random House, 2000).
Mithen, Steven, *Prehistory of the Mind: The Cognitive Origins of Art, Religion and Science* (London: Thames and Hudson, 1996).
Montagu, Ashley, "Review of *Wolf Children and Feral Man* by J.A.L. Singh and Robert M. Zingg," *American Anthropologist* 45, No. 3 (1943): 468–472.
Montagu, Ashley, *The Natural Superiority of Women* (New York: Macmillan, 1953).
Montagu, Ashley, *The Idea of the Primitive* (New York: The Free Press, 1968).
Montagu, Ashley, *Man's Most Dangerous Myth: The Fallacy of Race* (New York: Meridian Books, 1964).
Montagu, Ashley, and Edward Darling, *The Ignorance of Certainty* (New York: Harper and Row, 1970).
Moody, Paul Amos, *Genetics of Man* (New York: W.W. Norton, 1967).
Moon, Seok Young, "Cultural perspectives on attention-deficit hyperactivity disorder: a comparison between Korea and the U.S.," *Journal of International Business and Cultural Studies* 6 (2012): 1–11.
Moreno, Eduardo Lopez, and Rasna Warah, "The State of the World's Cities Report 2006/7: urban and slum trends in the 21st century," *UN Chronicle* 43, No. 2 (2006).
Morgan, L.H., *American Beaver and His Works* (New York: Dover, 1986). Originally published in 1868.
Morris, Brian, *Anthropological Studies of Religion* (Cambridge: Cambridge University Press, 1987).
Morris, Desmond, *The Human Zoo* (New York: Dell Books, 1969).
Morwood, M.J., P. Brown, Jatmiko et al., "Further evidence for small-bodied hominins from the Late Pleistocene of Flores, Indonesia" *Nature* 437, No. 7061 (October 13, 2005): 1012–1017.
Moyo, Dambisa, "Why foreign aid is hurting Africa," *Wall Street Journal*, March 21, 2009.
Mueller, Ulrich, "Is there a stabilizing selection around average fertility in modern human populations?" *Population and Development Review* 27, No. 3 (September 2001): 469–498.
Mullen, Sydney L.W., *The Evolution of Love* (San Francisco: W.H. Freeman, 1981).
Muller, Max F., *Natural Religion; the Gifford Lectures Delivered Before the University of Glasgow in 1888,* 2nd edition (London: Longmans, Green, 1892).
Muri, Gabriela, Pause!: Zeitordunug und Austeiten aus alltagskultureller Sicht (Frankfurt: Campus Verlag; Auflage: 1, November 10, 2004).
Murry, Caitlin, "The 'Colouring of the Psychosis': interpreting insanity in the primitive mind," *Health and History* 9, No. 2, "Aboriginal health and history" (2007): 7–21.

Mysore, Shreesh, and Eric I. Knudsen, "Descending control of neural bias and selectivity in a spatial attention network: rules and mechanisms," *Neuron* 84, No. 1 (October 1, 2014): 214–226.

Napier, John, and Russell H. Tuttle, *Hands* (Princeton: Princeton Science Library, Princeton University Press, 1993).

Naroll, Raoul, "Floor area and settlement population," *American Antiquity* 27, No. 4 (1962): 587–589.

Nathan, Carl F., *Plague Prevention and Politics in Manchuria: 1910–1931* (Cambridge: Harvard University Press, 1967).

National Association of Counties, *The Meth Epidemic in America II: Two New Surveys of U.S. Counties* (Washington, DC: National Association of Counties, 2006).

Needham, Joseph, *Science in Traditional China* (Cambridge: Harvard University Press, 1981).

Neel, James V., "Lessons from a 'primitive people,'" *Science* 170, No. 3960 (November 20, 1970): 815–822.

Neisser, Ulrich, *Cognition and Reality: Principles and Implications of Cognitive Psychology* (New York: Freeman, 1976).

Nelkin, D., and S.L. Gilman, "Placing blame for devastating disease," in *In Time of Plague: The History and Social Consequences of Lethal Epidemic Disease*, edited by A. Mack (New York: New York University Press, 1991): 39–56.

Nelson, C. Mindy, Kate E. Ihle, M. Kim Fondrk, Robert E. Page, Jr., and Fro V. Amdam, "The gene vitellogenin has multiple coordinating effects on social organization," *PLoS Biology* 5, No. 3 (2007): e62.

Neuwirth, Robert, *Shadow Cities: A Billion Squatters, A New Urban World* (London: Routledge, 2006).

New York City Department of Homeless Services Report, March 4, 2009.

Niebuhr, H. Richard, *The Social Sources of Denominationalism* (New York: World, 1957). Originally published in 1929.

Nieh, James C., "A negative feedback signal that is triggered by peril curbs honey bee recruitment," *Current Biology* 20 (2010): 310–315.

Nietzsche, F., *The Twilight of the Idols and the Anti-Christ: Or How to Philosophize with a Hammer*, translated by R.J. Hollingdale (New York: Penguin Books, 1991). Written in 1888, originally published in 1889.

Nieuwenhuys, R., H.J. Ten Donkelaar, and C. Nicholson, *The Central Nervous System of Vertebrates*, vol. 3 (Berlin: Springer, 1998).

Nissimi, Hilda, "Religious conversion, covert defiance and social identity: a comparative view," *Numen* 51, No. 4 (2004): 367–406.

Noam, Eli, *Media Ownership and Concentration in America* (Oxford: Oxford University Press, 2009).

Norwood, Graham, "Rise of the 'ghost homes'," *Financial Times, House and Home,* February 13, 2016.

Noy, D. "When framing fails: Ideas, influence, and resources in San Francisco's homeless policy field. *Social Problems* 56, No. 2 (2009): 223–242.

Oberheim, N., X. Wang, S. Goldman, and S.A. Nedergaard, "Astrocytic complexity distinguishes the human brain," *Trends in Neurosciences* 29 (2006): 1–10.

Oberheim, N., T. Takano, X. Han et al., "Uniquely hominid features of adult human astrocytes," *Journal of Neuroscience* 29 (2009): 3276–3287.

O'Brien, John E., and E. Lois Sampson, "Lipid composition of the normal human brain: gray matter, white matter, and myelin," *Journal of Lipid Research* 6 (1965): 537–544.

O'Conner, Sarah, "Tent cities pose challenges for US census," *Financial Times*, April 3, 2009.

O'Donnell, Sean, Marie Clifford, Yamile Molina, and Gene E. Robinson, "Comparative analysis of constraints and caste differences in brain investment among social paper wasps," *Proceedings of the National Academy of Sciences, USA* 108, No. 17 (April 26, 2011): 7107–7112.

Okasha, Samir, "Recent work on the levels of selection problem," *The Human Nature Review* 3 (2003): 349–356.

Okasha, S., *Evolution and the Levels of Selection* (Oxford: Oxford Univesity Press, 2006).

Oldroyd, B.P., "Evaluation of Australian commercial honey bees for hygienic behavior, a critical character for tolerance to chalkbrood," *Australian Journal of Experimental Agriculture* 36, No. 5 (1996): 635–629.

Olkowicz, Severyn, Martin Kocourek, Radek K. Lucan et al., "Birds have primate-like numbers of neurons in the forebrain," *Proceedings of the National Academy of Sciences, USA* 113, No. 26 (June 28, 2016): 7255–7260.

O'Neil, Cathy, *Weapons of Math Destruction: How Big Data Increases Inequality and Threatens Democracy* (New York: Allen Lane, 2016).

Opie, Christopher, Quentin D. Atkinson, Robin I.M. Dunbar, and Susanne Shultz, "Male infanticide leads to social monogamy in primates," *Proceedings of the National Academy of Sciences, USA* 110, No. 33 (August 13, 2013): 13328–13332.

O'Riain, M. Justin, and Chris G. Faulkes, "African mole rats: eusociality, relatedness and ecological constraints", in *Ecology of Social Evolution*, edited by Judith Korb and Jörgen Heinze (Berlin: Springer, 2008): 207–223.

Ornoy, A., and Z. Ergaz, "Alcohol abuse in pregnant women: effects on the fetus and newborns mode of action and maternal treatment," *International Journal of Environmental Research and Public Health* 7, No. 2 (2010): 364–379.

Orr, Robert T., *Vertebrate Biology* (Philadelphia: W.B. Saunders, 1966).

Ortner, Donald J., and Walter G.J. Putschar, *Identification of Pathological Conditions in Human Skeletal Remains* (Washington, DC: Smithsonian Institution Press, 1985).

Osborne, Henry Fairfield, *Man Rises to Parnassus: Critical Epochs in the Prehistory of Man* (Princeton: Princeton University Press, 1927).

Osborne, Henry Fairfield, *Men of the Old Stone Age* (New York: Scribners, 1916).

Oxford Universal Dictionary, 1955 edition (Oxford: Oxford University Press, 1955).

Pagel, Mark D., and Paul H. Harvey, "How mammals produce large-brained off-spring," *Evolution* 42 (1988): 948–957.

PAHO and WHO, "Increase in microcephaly in the Northeast of Brazil," Epidemiological Report, November 17, 2015.

Palombo, Maria Rita, S. Moya-Sola, M. Kohler, and C. Giovinazzo, "Brain versus body mass in endemic ruminant artiodactyls: a case studied of *Myotragus balearicus* and smallest *Candiacervus* species from Mediterranean islands," *Quaternary International* 182 (September 2007): 160–183.

Palumbi, Stephen R., *The Evolution Explosion* (New York: W.W. Norton, 2001).
Panksepp, Jaak, "The riddle of laughter neural and psychoevolutionary underpinnings of joy," *Current Directions in Psychological Science* 6, No. 9 (December 2000): 183–186.
Panksepp, Jaak, "Neurologizing the psychology of affects: how appraisal-based constructivism and basic emotion theory can coexist," *Perspectives on Psychological Science* 2, No. 3 (September 2007): 281–296.
Paredes, Mercedes, Shawn F. Sorrells, Jose M. Garcia-Verdugo, and Arturo Alvarez-Buylla, "Brain size and limits to adult neurogenesis," *Journal of Comparative Neurology* 524 (2016): 646–664.
Park, Robert E., *Human Communities: The City and Human Ecology* (Glencoe, IL: The Free Press, 1952).
Park, Robert E., E.W. Burgess, and R.D. McKenzie, *The City* (Chicago: University of Chicago Press, 1967).
Parker, S.T., "A sexual selection model for hominid evolution," *Human Evolution* 2, No. 3 (1987): 235–253.
Parker, S.T., and K.R. Gibson, "A developmental model for the evolution of language and intelligence in early hominids," *Behavioral and Brain Sciences* 2 (1979): 367–408.
Parker, S.T., and Michael L. Mckinney, *Origins of Intelligence* (Baltimore: Johns Hopkins Press, 1999).
Passaro, Joanne, *The Unequal Homeless: Men on the Streets, Women in Their Place* (New York: Routledge, 1996).
Patterson, Francine, and E. Linden, *The Education of KoKo* (New York: Henry Holt, 1981).
Pearce, Eiluned, Chris Stringer, and R.I.M. Dunbar, "New insights into differences in brain organization between Neanderthals and anatomically modern humans," *Proceedings of the Royal Society B* (March 13, 2013). doi: 10.1098/rspb.2013.0168.
Pearce, J.M.S., "Marie-Jean-Pierre Flourens and (1794–1867) and cortical localization," *European Neurology* 61 (2009): 311–314.
Peirce, Charles Sanders, "Logic as semiotic: the theory of signs," in *The Philosophical Writings of Peirce*, edited by J. Buchler (New York: Dover Books, 2011): 98–119.
Pellens, Roseli, Frederic Legendre, and Philippe Gradcolas, "Phlylogenetic analysis of social behavior evolution in (*Zetoborinae* and *Blaberinae* and *Byninae* and *Diplopterinae*) cockroaches: an update with the study of endemic radiations from the Atlantic forest," *Studies on Neotropical Fauna and Environment* 42, No. 1 (April 2007): 25–31.
Pena-Melian, A., "Development of human brain," *Human Evolution* 15, No. 1–2 (2000): 99–112.
Pépin, Jacques, *The Origins of AIDS* (Cambridge: Cambridge University Press, 2011).
Perham, N., H. Hodgetts, and S. Banbury, "Mental arithmetic and non-speech office noise: an exploration of interference-by-content," *Noise Health* 15, No. 62 (February 2013): 73–78.
Perner, Josef, and Martin Doherty, "Do infants understand that external goals are internally represented?" *Behavioral and Brain Sciences* 28, No. 5 (2005): 710–711.

Perou, Ruth, Rebecca Bitsko, Steven J. Blumberg et al., "Mental health surveillance among children, 2005–2011," *CDC Supplements* 62, No. 2 (May 2013): 1–35.

Perucho, Manuel, "Microsatellite instability: the mutator that mutates the other mutator," *Nature Medicine* 2, No. 6 (1996): 630–631.

Petro, Allison, "The language instinct, reviewed," *TESOL Quarterly* 30, No. 2 (Summer 1996): 365–366.

Petronius, *Satyricon*, translated by Michael Heseltine (London: Heinmann, 1913).

Pettifor, Philiip, "The decline of the public library," *Spiked*, September 2, 2005.

Pew Research Center, "Parenting in America," *Pew Surveys*, December 17, 2015.

Pfeiffer, J.E., *The Creative Explosion: An Inquiry Into the Origins of Art and Religion* (New York: Harper Collins, 1982).

Pfeifer, R., and J. Bongard, *How the Body Shapes the Way We Think: A New View of Intelligence* (Cambridge: MIT Press, 2007).

Piaget, J., *Play, Dreams and Imitation in Childhood* (New York: W.W. Norton, 1962). Originally published in 1951.

Pilling, David, "Megacities," *Financial Times*, November 4, 2011.

Pinker, Steven, *Learnability and Cognition: The Acquisition of Structure* (Cambridge: MIT Press, 1989).

Pinker, Steven, *The Language Instinct* (New York: William Morrow, 1994).

Pinker, Steven, *How the Mind Works* (New York: Norton, 1997).

Pinker, Steven, *The Blank Slate* (New York: Viking Books, 2002).

Pinker, Steven, *The Better Angels of Our Nature: Why Violence Has Declined* (New York: Viking Books, 2011).

Plomin, Robert, Michael Owen, and Peter McGuffin, "The genetic basis of complex human behaviors," *Science* 264 (June 17, 1994): 1733–1739.

Plutarch, *PLUTARCH: Lives of the noble Grecians and Romans*, translated and edited by A.H. Clough (New York: A.L. Burt, 1910). Originally published in 1859.

Polanyi, Karl, *The Great Transformation* (New York: Holt, Rinehart and Winston, 1944).

Pollio, H.R., *The Psychology of Symbolic Activity* (Reading, MA: Addison-Wesley, 1974).

Pollock, Naomi, *Modern Japanese House* (London: Phaidon, 2006).

Polybius, 200–118 B.C.E. *The Rise of the Roman Empire* (London: Penguin, 1979).

Powell, Jane, "Green envy: what are the true environmental benefits of the 'Green' house movement?" *San Francisco Chronicle Magazine*, May 13, 2007.

Premack, D., "Human and animal cognition: continuity and discontinuity," *Proceedings of the National Academy of Sciences, USA* 104 (2007): 13861–13867.

Preuss, T.M., "The discovery of cerebral diversity: an unwelcome scientific revolution," in *Evolutionary Anatomy of the Primate Cerebral Cortex*, edited by D. Falk and K. Gibson (Cambridge: Cambridge University Press, 2001): 138–164.

Price, E.O., *Principles and Applications of Domestic Animal Behavior* (Wallingford, UK.: CABI, 2008).

Prusiner, Stanley B., "Neurodegenerative diseases and prions," *New England Journal of Medicine* 344 (May 17, 2001): 1516–1526.

Puigbo, Pere, Yuri I. Wolf, and Eugene V. Koonin, "Search for a 'Tree of Life' in the thicket of the phylogenetic forest," *Journal of Biology* 8, No. 59 (2009): 1–17.

Quine, W.V.O., *Word and Object* (Cambridge: MIT Press, 1960).

Radin, Paul, *Primitive Religion* (New York: Dover Press, 1957). Originally published in 1937.

Rambo, Louis R., *Understanding Religious Conversion* (New Haven: Yale University Press, 1993).

Ratcliffe, J.M., M.B. Fenton, and S.J. Shettleworth, "Behavioral flexibility positively correlated with relative brain volume in predatory bats," *Brain, Behavior and Evolution* 67, No. 3 (2006): 165–176.

Radcliffe-Brown, A.R., *The Andaman Islanders: A Study in Social Anthropology* (Cambridge: Cambridge University Press, 1922).

Ratzel, Friedrich, *The History of Mankind* (London: Macmillan, 1896).

Reber, A.S., "The cognitive unconscious: an evolutionary perspective," *Consciousness and Cognition* 1 (1992): 93–133.

Reber, A.S., "Caterpillars, consciousness and the origins of mind," *Animal Sentience* 1, No. 11 (2016).

Redfield, Robert, *Folk Culture of Yucatan* (Chicago: University of Chicago Press, 1941).

Redfield, Robert, and Milton B. Singer, "The cultural role of cities," *Economic Development and Cultural Change* 3, No. 1 (October 1954): 53–73.

Redman, Charles L., *Human Impact on Ancient Environments* (Tucson: University of Arizona Press, 1999).

Regen, A., L.P. Nelson, and S.B. Woo, "Dental manifestations associated with Seckel syndrome type II: a case report," *Pediatric Dentistry* 32, No. 5 (September–October 2010): 445–450.

Reeve, D., "Psycho-emotional disablism in the lives of people experiencing mental distress," in *Madness, Distress and the Politics of Disablement,* edited by H. Spandler, J. Anderson, and B. Sapey (Bristol: Policy Press, 2015a): 99–112.

Reeve, D., "Disgust, self-disgust and disability: a disability study in perspective," in *The Revolting Self: Perspectives on the Psychological and Clinical Implications of Self-Directed Disgust,* edited by P. Powell, P. Overton, and J. Simpson (London: Karmac Books, 2015b): 53–74.

Reeve, John Marshall, *Understanding Motivation and Emotion* (Hoboken: Wiley, 2015).

Reischauer, E.O., and J.K. Fairbank, *East Asia: The Great Tradition* (Boston: Houghton Mifflin, 1958).

Reiter, Reyna R., editor, *Toward an Anthropology of Women* (New York: Monthly Review Press, 1975).

Rescorla, R.A., "Effect of US habituation following conditioning," *Journal of Comparative Physiology* 82 (1973): 137–143.

Reuter, M., J.S. Peterson, and L. Keller, "Loss of Wolbachia infection during colonisation in the invasive Argentine ant Linepithema humile," *Heredity* 94 (2005): 364–369.

Richardson, Sara S., "Race and IQ in the postgenomic age: the microcephaly case," *BioSocieties* 6 (2011): 420–446. doi: 10.1057/biosoc.2011.20, published online October 17, 2011.

Riley, J.R., U. Greggers, A.D. Smith, D.R. Reynolds, and R. Menzel, "The flight paths of honeybees recruited by the waggle dance," *Nature* 435 (May 12, 2005): 205–207.

Ris, Hans, "Primitive mitotic mechanisms," *Biosystems* 7 (1975): 298–304.

Riveros, Andre, Marc A. Seid, and William T. Wcislo, "Evolution of brain size in class-based societies of fungus-growing ants (Attini)," *Animal Behavior* 83 (2012): 1043–1049.

Roebroeks, W., and P. Villa, "On the earliest evidence for habitual use of fire in Europe," *Proceedings of the National Academy of Sciences, USA* 108 (2011): 5209–5214.

Roeder, Kenneth D., "Episodes in insect brains: how do animals select modes of behavior adapted to immediate circumstances? A mechanism is sought by probing the brains of insects," *American Scientist* 58, No. 4 (1970): 378–389.

Rogers, L.J., and R.J. Andrews, editors, *Comparative Vertebrate Lateralization* (Cambridge: Cambridge University Press, 2002).

Roheim, Geza, *The Eternal Ones of the Dream* (New York: International Universities Press, 1945).

Roisin, Yves, and Judith Korb, "Social organization and the status of workers in termites," in *Biology of Termites: A Modern Synthesis*, edited by David Edward Bignell, Yves Roisin, and Nathan Lo (New York: Springer, 2011): 133–164.

Rosengren, R., and W. Fortelius, "Ortstreue in foraging ants of the Formica rufa group – hierarchy of oriented cues and long-term memory," *Insectes Sociaux* 33 (1986): 3060337.

Rosenthal, Franz, "The stranger in modern Islam," *Arabica*, 44, No. 1 (January, 1997): 35–75.

Rossignol-Strick, Martine, "The Holocene climatic optimum and pollen records of sapropel 1 in the eastern Mediterranean, 9000–6000 BP," *Quaternary Science Reviews* 18, No. 4–5 (April 1999): 515–530.

Ross, Marina Davila, Michael J. Owren, and Elke Zimmermann, "Reconstructing the evolution of laughter in Great Apes and humans," *Current Biology* 19, No. 13 (July 14, 2009): 1106–1111.

Rostovtzeff, M.I., *Social and Economic History of the Roman Empire* (Oxford: Oxford University Press, 1926).

Roth, G., and U. Dicke, "Evolution of the brain and intelligence," *Trends in Cognitive Sciences* (regular edition) 9, No. 5 (May 2005): 250–257.

Rothner, A.D., and A. Gupta, "Treatment of childhood headaches," *Current Neurology and Neuroscience Reports* 1, No. 2 (March 2001): 144–154.

Rothenbuhler, W.C., "Behavioral genetics of nest cleaning in honey bees, IV, responses of F1 and backcross generations to disease-killed brood," *American Zoologist* 4 (1964): 111–123.

Rouseau, Phillip, *Ascetics, Authority and the Church in the Age of Jerome and Cassan* (Oxford: Oxford University Press, 1978).

Rowe, John Howland, "Worsaae's Law and the use of grave lots for archaeological dating," *American Antiquity* 28, No. 2 (October 1962): 129–137.

Rhodes, Richard, *Deadly Feasts: "The Prion" Controversty and the Public's Health* (New York: Touchstone Books, 1998).

Rupert, Robert D., "Coining terms in the language of thought: innateness, emergence, and the lot of Cummins's argument against the causal theory of mental content," *Journal of Philosophy* 98, No. 10 (October 2001): 499–530.

Russell, Paul, "Skepticism and natural religion in Hume's Treatise," *Journal of the History of Ideas* 49, No. 2 (April–June 1988): 247–265.

Russell, Sabin, "Lethal new bacteria—S.F. an epicenter," *San Francisco Chronicle*, January 15, 2008.

Ryan, Joan, "Medicine to console the sullen, anti-depressants can calm, uplift troubled teenagers," *San Francisco Chronicle*, March, 23, 2004.

Rybczynski, Natalia, "Woodcutting behavior in beavers (Castoridae, Rodentia): estimating ecological performance in a modern and a fossil taxon," *Paleobiology* 34, No. 3 (Summer 2008): 389–402.

Sabatini, J. (2014). SF's latest housing problem: Evictions of formerly homeless residents. San Francisco Examiner, 21 February.

Sabloff, Jeremy, "Settlement patterns and community organization in the Maya Lowlands" *Expedition Magazine* 38, No. 1 (March 1996). http://www.penn.museum/sites/expedition/?p=4895, accessed January 5, 2017.

Sacks, Oliver, *The Man Who Mistook His Wife for a Hat* (New York: Harper and Row, 1970).

Sagan, Carl, and I.S. Shklovskiĭ, *Intelligent Life in the Universe* (San Francisco: Holden-Day, 1966).

Sakai, Tomoko, Satoshi Hirata, Kohki Fuwa et al., "Fetal brain development in chimpanzees versus humans," *Current Biology* 22, No. 18 (September 25, 2012): R791–R792.

Sallet, J., R.B. Mars, M.P. Noonan et al., "Social network size affects neural circuits in macaques," *Science* 334 (2011): 697–700.

Sampson, G., *The "Language Instinct" Debate,* revised (New York: Continuum Press, 2005).

Santos, Andreia, Andreas Meyer-Lindenberg, and Christine Deruelle, "Absence of racial, but not gender, stereotyping in Williams syndrome children," *Current Biology* 20, No. 7 (April 13, 2010): R307–R308.

Sassen, Saskia, *Cities in a World Economy* (Thousand Oaks, CA: Pine Forge Press, 2012).

Saussure, de Ferdinand, Cours de linguistique générale, edited by C. Bally and A. Sechehaye with the collaboration of A. Riedlinger (Lausanne and Paris: Payot, 1916). Published in English as *Course in General Linguistics*, translated by W. Baskin (Glasgow: Fontana/Collins, 1977).

Savage-Rumbaugh, E.S., *Ape Language: From Conditioned Response to Symbol* (New York: Columbia University Press, 1986).

Scheper-Hughes, Nancy "The global traffic in human organs," *Current Anthropology* 41, No. 2 (2000): 1–59.

Schick, Kathy, and Nicholas Toth, *Making Silent Stones Speak* (New York: Simon and Schuster, 1993).

Schoenemann, P.T., "Evolution of the size and functional areas of the human brain," *Annual Review of Anthropology* 35 (2006): 379–406.

Schoenemann, Thomas P., Thomas F. Budinger, Vincent M. Sarich, and William S.-Y. Wang, "Brain size does not predict general cognitive ability within families," *Proceedings of the National Academy of Sciences, USA* 97, No. 9 (April 25, 2000): 4932–4937.

Schreiweiss, Christine, Ulrich Bornschein, Eric Burguiere et al., "Humanized Foxp2 accelerates learning by enhancing transitions from declarative to procedural performance," *Proceedings of the National Academy of Sciences, USA* 111, No. 39 (2014): 14253–14258.

Schroeder, Lauren, Jill E. Scott, Heather M. Garvin et al., "Skull diversity in the Homo lineage and the relative position of *Homo nadeli*," *Journal of Human Evolution* 97 (August 2016): 17–26.

Schuessler, Gretchen, *Community Participation in Parks Planning: Addressing Environmental Degradation on Providence's South Side*, Ph.D. thesis, Center for Environmental Studies, Brown University, 1992.

Schultz, Ted R., and Sean G. Brady, "Major evolutionary transitions in ant agriculture," *Proceedings of the National Academy of Sciences, USA* 105, No. 14 (April 8, 2008): 5435–5440.

Schumpeter, J.A., *Business Cycles: A Theoretical, Historical and Statistical Analysis of the Capitalist Process* (New York: McGraw-Hill, 1939).

Schuster, Richard, "Why not chimpanzees, lions and hyenas, too," *Behavioral and Brain Sciences* 28, No. 5 (2005): 716–717.

Schwartz, Jeffrey H., and Ian Tattersall, "Defining the genus Homo," *Science* 349, No. 6251 (August 28, 2015): 931–932.

Scott, Erin, Janet Mann, Jana Watson-Capps, Brooke Sargeant, and Richard Connor, "Aggression in bottlenose dolphins: evidence for sexual coercion, male–male competition, and female tolerance through analysis of tooth-rake marks and behavior," *Behaviour* 142, No. 1 (2005): 21–44.

Scull, Andrew, *Madness in Civilization* (Princeton: Princeton University Press, 2015).

Seckel, Helmut P.G., *Bird-Headed Dwarfs: Studies in Developmental Anthropology Including Human Proportions* (Springfield: Charles C. Thomas, 1960).

Searle, John R., "Animal minds," *Midwestern Studies in Philosophy* XIX (1994): 206–219.

Seel, Norbert M., *Encyclopedia of the Sciences of Learning* (New York: Springer, 2012).

Seeley, T.D., "Decision making in superorganisms: how collective wisdom arises from the poorly informed masses," in *Bounded Rationality: The Adaptive Toolbox*, edited by G. Gigerenzer and R. Selten (Cambridge: MIT Press, 2001): 249–261.

Segal, Robert A., "Jung and Levy-Bruhl," *Journal of Analytical Psychology* 52 (2007): 635–658.

Self, Stephen, and Stephen Blake, "Consequences of explosive supereruptions," *Elements* 4, No. 1 (February 2008): 41–46.

Sekar, Aswin, Allison R. Bialas, Heather de Rivera et al., "Schizophrenia risk from complex variation of complement component 4," *Nature* 530 (February 11, 2016). doi: 10.1038/nature16549.

Semple, Bridgette D., Klas Blomgren, Kayleen Gimlin, Donna M. Ferriero, and Linda J. Noble-Haeusslein, "Brain development in rodents and humans: identifying

benchmarks of maturation and vulnerability to injury across species," *Progress in Neurology* 106, No. 107 (2013): 1–16.
Sender, R., S. Fuchs, and R. Milo, "Are we really vastly outnumbered? Revisiting the ratio of bacterial to host cells in humans," *Cell* 164, No. 3 (2016): 337–340. doi: 10.1016/j.cell.2016.01.013.
Serre, David, and Svante Paabo, "Evidence for gradients of human genetic diversity within and among continents," *Genome Research* 14 (2004): 1679–1685.
Shariff, G.A. "Cell counts in the primate cerebral cortex," *Journal of Comparative Neurology* 98 (1953): 382–400.
Sharma, Arvind, *Hinduism as a Missionary Religion* (Albany: SUNY Press, 2011).
Shaw, K., I. Wagner, H. Eastwood, and G. Mitchell, "A qualitative study of ADHD in Australia," *Australia Family Practice* 20, No. 2 (2003): 129–134. doi: 10.1093/fampra/20.2.129.
Sherwood, Chet C., Cheryl D. Stimpson, Mary Ann Raghanti et al., "Evolution of increased glia-neuron ratios in the human frontal cortex," *Proceedings of the National Academy of Sciences, USA* 103, No. 37 (September 12, 2006): 13606–13611.
Shiffrin, Richard, and Robert Nosofsky, "Seven plus or minus two: a commentary on capacity limitations," *Psychological Review* 101 (1994): 357–361.
Shreeve, Jaime, "Oldest human fossil found, rewriting family tree," *National Geographic Magazine*, March 5, 2015a.
Shreeve, Jaime, "This face changes the human story, but how?" *National Geographic Magazine,* September 10, 2015b.
Shik, Jonathan Z., Chen Hou, Adam Kay et al., "Towards a general life-history model of the superorganism: predicting the survival, growth and reproduction of ant societies," *Biology Letters* 8, No. 6 (December 23, 2012): 1059–1062.
Shomrat, T., A.L. Turchetti-Maia, N. Stem-Mentch, J.A. Basil, and B. Hochner, "The vertical lobe of cephalopods: an attractive brain structure for understanding the evolution of advanced learning and memory systems," *Journal of Comparative Physiology A* 201, No. 9 (September 2015): 947–956.
Sieverts, Thomas, *Cities without Cities: An Interpretation of the Zwischenstadt* (London: Spon Press, 2003).
Sigerist, Henry E., *A History of Medicine* (New York: Oxford University Press, 1987).
Silva Jr., J., F. Silva, and I. Sazima, "Rest, nurture, sex, release, and play: diurnal underwater behaviour of the spinner dolphin at Fernando de Noronha Archipelago, SW Atlantic – mating behaviour," *Aqua, Journal of Ichthyology and Aquatic Biology* 9 (2005): 161–176.
Silverman, Paul H., "Rethinking genetic determinism," *The Scientist* (May 24, 2004): 32–33.
Simpson, G.G., *The Meaning of Evolution* (New Haven: Yale University Press, 1949).
Singh, A.L., and Robert M. Zingg, *Wolf-Children and Feral Man* (New York: Harper, 1942).
Singh, Ilina, "Doing their jobs: mothering with Ritalin in a culture of mother-blame," *Social Science and Medicine* 59, No. 6 (2004): 1193–1205.

Sjoberg, Gideon, *The Preindustrial City: Past and Present* (New York: Free Press, 1960).

Skinner, B.F., *Science and Human Behavior* (New York: The Free Press, 1953).

Skinner, B.F., "Superstition in the pigeon," *Journal of Experimental Psychology* 38 (1948): 168–172.

Skouri, H., R. Gandouz, J. Kraiem et al., "Antibodies to anionic phospholipids and cofactors in kala-azar. Comparative study with malaria, toxoplasmosis and 'autoimmune' diseases," *Clinical and Experimental Rheumatology* 5 (September–October 2008): 894–902.

Slobodchikoff, C.N., Bianca S. Perla, and Jennifer L. Verdolin, *Prairie Dogs: Communication and Community in an Animal Society* (Cambridge: Harvard University Press, 2009).

Smaers, Jeroen B., Dina K.N. Dechmann, Anjali Goswami, Christophe Soligo, and Kamran Safi, "Comparative analyses of evolutionary rates reveal different pathways to encephalization in bats, carnivorans, and primates," *Proceedings of the National Academy of Sciences, USA* 109, No. 44 (October 30, 2012): 1806–18011.

Smaers, Jeroen, and Christophe Soligo, "Brain reorganization, not relative brain size, primarily characterizes anthropoid brain evolution," *Proceedings, Biological Sciences* 280, No. 1759 (May 22, 2013): 1–8.

Smil, Vaclav, *The Earth's Biosphere: Evolution, Dynamics and Change* (Cambridge: MIT Press, 2003).

Smith, Adam, *The Wealth of Nations* (New York: Arlington House, 1965). Originally published in 1776.

Smith, A.H.V., "Provenance of coals from Roman sites in England and Wales," *Britannia* 28 (1997): 297–324 (322–324).

Smith, Debra R.R., "Population genetics of *Anelosimus eximius* (Araneae: Theridiidae)," *Journal of Arachnology* 14 (1986): 201–217.

Smith, Jennifer E., Eli M. Swanson, Daphna Reed, and Kay E. Holekamp, "Evolution of cooperation among mammalian carnivores and its relevance to Hominin evolution," *Current Anthropology* 53, No. S6, "Human Biology and the Origins of Homo" (December 1, 2012): S436–S452.

Smith, Moira, "Humor, unlaughter and boundary maintenance," *Journal of American Folklore* 122, No. 2 (2009): 148–171.

Smulders, T.V., M.W. Shiflett, A.J. Sperling, and T.J. DeVoogd, "Seasonal changes in neuron numbers in the hippocampal formation of a food-hoarding bird: the black-capped chickadee," *Journal of Neurobiology* 44, No. 4 (September 15, 2000): 414–422.

Sojberg, Gideon, *The Preindustrial City* (New York: The Free Press, 1965).

Sol, Daniel, "Revisiting the cognitive buffer hypothesis for the evolution of large brains," *Biology Letters*, February 23, 2009.

Sowell, E.R., P.M. Thompson, S.E. Welcome et al., "Cortical abnormalities in children and adolescents with attention-deficit hyperactivity disorder," *Lancet* 362, No. 22 (November 2003): 1699–1706.

Spearman, C. "'General intelligence,' objectively determined and measured," *The American Journal of Psychology* 15, No. 2 (1904): 201–292.

Spencer, Herbert, *The Principles of Sociology*, vol. 1 (New York: D. Appleton, 1898). http://oll.libertyfund.org/titles/2642, accessed January 6, 2017.

Spierenburg, Pieter, *A History of Murder: Personal Violence in Europe from the Middle Ages to the Present* (Cambridge: Polity Press, 2008).
Spiro, Melford E., *Buddhism and Society,* 2nd edition (Berkeley: University of California Press, 1982).
Spivak, Polma, and Dani Nadel, "The use of stone at Ohalo II, a 23,000 year old site in the Jordan Valley, Israel," *Journal of Lithic Studies* 3, No. 3 (2013): 523–552.
Srinivasan, Mandyam V., "Honey bees as a model for vision, perception, and cognition," *Annual Review of Entomology* 55 (January 2010): 267–284.
Stampe, D., "Toward a causal theory of linguistic representation," in *Midwest Studies in Philosophy*, vol. 2, edited by P. French, H.K. Wettstein, and T.E. Uehling (Minneapolis: University of Minnesota Press, 1977): 42–63.
Stanford, C.B., "The hunting ecology of wild chimpanzees: implications for the evolutionary ecology of Pliocene hominids," *American Anthropologist* 98, No. 1 (1996): 96–113.
Stanhope, Michael J., Andrei Lupas, Michael J. Italia, Kristin K. Koretka, Craig Volker, and James R. Brown, "Phylogenetic analyses do not support horizontal gene transfers from bacteria to vertebrates," *Nature* 411, No. 21 (June 2001): 940–944.
Stark, Miriam T., "From Funan to Angkor: collapse and regeneration in Ancient Cambodia," in *After Collapse: The Regeneration of Complex Societies*, edited by Glenn M. Schwartz and John J. Nichols (Tucson: University of Arizona Press, 2006): 144–167.
Starr, Kevin, *Inventing the Dream* (Oxford: Oxford University Press, 1985).
Stausfeld, Nicholas J., Lars Hansen, Yongsheng Li, Robert S. Gomez, and Kei Ito, "Evolution, discovery and interpretations of Arthropod Mushroom Bodies," *Learning and Memory* 5 (1998): 11–37.
Steadman, Sharon R., "Heading home: the architecture of family and society in early sedentary communities on the Anatolian Plateau," *Journal of Anthropological Research* 60, No. 4 (Winter 2004): 515–558.
Steinbock, R. Ted, *Paleopathological Diagnosis and Interpretation* (Springfield: Charles C. Thomas, 1976).
Steklis, Hieter H., "The Primate brain and the origins of intelligence," in *The New Physical Anthropology*, edited by Shirley C. Strum, Donald G. Lindburg, and David Hamburg (Upper Saddle River: Prentice Hall, 1999): 47–63.
Sternberg, R.J., *Beyond IQ: A Triarchic Theory of Human Intelligence* (New York: Cambridge University Press, 1985).
Sternberg, Robert J., "Beyond g: the theory of successful intelligence," in *The General Factor in Intelligence: How General Is It?*, edited by R.J. Sternberg and E.L. Grigorenko (Mahwah, NJ: Lawrence Erlbaum Associates, 2002).
Sternberg, Robert J., and Elena L. Grigorenko, "Intelligence and culture: how culture shapes what intelligence means, and the implications for a science of well-being," *The Royal Society, Philosophical Transaction: Biological Sciences* 359, No. 1449 (2004): 1427–1434.
Stettner, L.J., and K.A. Matyniak, "The brain of birds," *Scientific American* 218 (June 6, 1968): 64–77.

Stith, Sara S., and Jacob Vigil, "Does ultrasociality really exist—and is it the best predictor of human economic behaviors?" *Behavioral and Brain Sciences* 39 (2016): 36–37.

Stocking, George W., *Race, Culture and Evolution: Essays in the History of Anthropology* (New York: The Free Press, 1968).

Stone, Linda, *Kinship and Gender* (Boulder, CO: Westview Press, 2010).

Stout, Dietrich, Nicholas Toth, Kathy Schick, and Thierry Charminade, "Neural correlates of Early Stone Age toolmaking: technology, language and cognition in human evolution," *Philosophical Transactions of the Royal Society of London B, Biological Sciences* 363, No. 1499 (June 12, 2008): 1939–1949.

Strachan, Mark W.J., Jacqueline F. Price, and Brian M. Frier, "Diabetes, cognitive impairment and dementia," *British Medical Journal* 336, No. 7634 (January 5, 2008): 6. doi: 10.1136/bmj.39386.664016.BE.

Strange, James, *James Strange's Journal and Narrative of the Commercial Expedition from Bombay to the North-West Coast of America Together with A Chart Showing the Tract of the Expedition*, introduction by A.V. Venkatarama Ayaar (Madras: Reprinted by the superintendent, Government Press, 1929).

Strasfeld, Nicholas J., Lars Hansen, Youngsheng Li, Robert S. Gomez, and Kei Ito, "Evolution, discovery and interpretations of arthropod mushroom bodies," *Learning and Memory* 5, No. 1–2 (May–June 1998): 11–37.

Striedter, G.F., *Principles of Brain Evolution* (Sunderland, MA: Sinauer Associates, 2005).

Striedinger, Ivo, "Neues Schrifttum uber Kaspar Hauser," *Zeitschrift fur bayerische Landesgeschichte* 6 (1933): 415–484.

Stringer, C.B., and P. Andrews, "Genetic and fossil evidence for the origin of modern humans," *Science* 239 (March 11, 1988): 1263–1269.

Strohle, A., M. Stoy, J. Wrase et al., "Reward anticipation and outcomes in adult males with attention deficit/hyperactivity disorder," *Neuroimage* 39, No. 3 (2008): 966–972.

Strong, William Duncan, "An introduction to Nebraska archaeology," *Smithsonian Miscellaneous Collections* 93, No. 10 (Washington, DC: Smithsonian Institution, 1935).

Studdert-Kennedy, Michael, and Louis Goldstein, "Launching language: the gestural origin of discrete infinity," in *Language Evolution*, edited by Morten H. Christiansen and Simon Kirby (Oxford: Oxford University Press, 2003). Published to Oxford Scholarship Online in 2010. doi: 10.1093/acprof:oso/9780199244843.003.0013.

Strupp, M., K. Hüfner, R. Sandmann et al., "Central oculomotor disturbances and nystagmus: a window into the brainstem and cerebellum," *Deutsches Arzteblatt International* 108, No. 12 (March 2011): 197–204. doi: 10.3238/arztebl.2011.0197.

Su, Songkun, Fang Cai, Aung Si, Shaowu Zhang, Jürgen Tautz, and Shenglu Chen, "East learns from West: Asiatic honeybees can understand dance language of European honeybees," *PLoS ONE* 3, No. 6 (June 4, 2008): e2365. doi: 10.1371/journal.pone.0002365.

Sudjic, Deyan, *The Edifice Complex: How the Rich and Powerful Shape the World* (London: Penguin Press, 2005).

Sullivan, Colin, "Human population growth creeps back up," *Scientific American* (June 14, 2013).

Sumbre, G., G. Florito, T. Flash, and B. Hochner, "Motor control of flexible octopus arms," *Nature* 433 (2005): 595–596.

Summer, Marianne, "Human tools of the European tertiary? Artifacts, brains and minds in evolutionist thinking, 1870–1920," *Notes and Records of the Royal Society of London* 65, No. 1 (March 20, 2011): 65–82.

Sun, Tao, and Robert F. Hevner, "Growth and folding of the mammalian cerebral cortex: from molecules to malformations," *Nature Reviews Neuroscience* 15 (April 2014): 217–232.

Susanka, Sarah, *The Not So Big House: A Blueprint for How We Really Live* (Newtown, CT: Taunton Press, 1998).

Swartz, Jeffrey H., and Bruno Maresca, "Do molecular clocks run at all? A critique of molecular systematic," *Biological Theory* 1, No. 4 (2006): 357–371.

Swedo, S.E., H.L. Leonard, and M.B. Shapiro, "Sydenham's Chorea: physical and psychological symptoms of St Vitus dance," *Pediatrics* 91, No. 4 (1993): 706–713.

Szatmari, P., "The epidemiology of attention-deficit hyperactivity disorders," *Child and Adolescent Psychiatric Clinics of North America* 1, No. 2 (October 1992): 361–371.

TangYa, Xie Jia-Sui, and Chen Keming, "Hand pollination of pears and its implications for biodiversity conservation and environmental protection: a case study from Hanyuan County, Sichuan Province, China," College of the Environment, Sichuan University, Moziqiao, Chengdu 610065, Sichuan China, 2001. https://buckfastimker.files.wordpress.com/2016/05/handbestuivingchina.pdf, accessed January 6, 2017.

Tainter, J.A., *The Collapse of Complex Societies* (Oxford: Oxford University Press, 1988).

Tanner, N., and A. Zihlman, "Women in evolution, Part I: innovation and selection in human origins," *Signs: Journal of Women in Culture and Society* 1 (1976): 585–608.

Teizer, Eva H., Kathryn L. Humphreys, Mor Shapiro, and Nim Tottenham, "Amygdala sensitivity to race is not present in childhood but emerges over adolescence," *Journal of Cognitive Neuroscience* 25, No. 2 (February 2013): 234–244.

Texier, Pierre-Jean, G. Porraz, J. Parkington et al., "A Howiesons Poort tradition of engraving ostrich egg shell containers dated to 60,000 years ago at Diepkloof Rock Shelter, South Africa," *Proceedings of the National Academy of Sciences, USA* 107, No. 14 (April 6, 2010): 6180–6185.

Thakur, Atul, "33% of Indians live in less space than US prisoners," *The Economic Times*, November 25, 2008.

Thapar, Romila, *Early India, From the Origins to 1300 A.D.* (Berkeley: University of California Press, 2002).

Thomas, Cyrus, *Introduction to the Study of North American Archaeology* (New York: AMS Press for the Peabody Museum of Archaeology and Ethnology, Harvard University, 1973). Originally published in 1898.

Thompson, Edward P., "Zeit, Arbeitsdisziplin und Industriekapitalismus," in *Plebeische Kultur und moralische Okonomie, Aufsatse zur englishcen Socialgeschachte des 18 und 19 Jahrhunderts* (Frankfurt: Ullstein, 1980): 34–66.

Thompson, Helen, "Oldest stone tools yet discovered are unearthed in Kenya," *The Smithsonian Magazine*, May 20, 2015.

Thompson, P.M., K.M. Hayashi, S.L. Simon et al., "Structural abnormalities in the brains of human subjects who use methamphetamine," *Journal of Neuroscience* 24, No. 26 (June 30, 2004): 6028–6036.

Thornton, Alex, and Katherine McAuliffe, "Teaching in wild meerkats," *Science* 313, No. 5784 (2006): 227–229.

Thucydides, *The Peloponnesian War/Thucydides; the Crawley Translation Revised, with an Introduction, by T.E. Wick* (New York: Modern Library, 1982). Written 431 B.C.E.

Thurstone, L.L., *Primary Mental Abilities* (Chicago: University of Chicago Press, 1957).

Tiedemann, Fredrich, "On the brain of the Negro, compared with that of the European and the orang-outang," *Philosophical Transactions of the Royal Society of London* 126 (1836): 497–527.

Tiger, Lionel, *Men in Groups* (New York: Random House, 1969).

Timimi, Sami, and Eric Taylor, "ADHD is best understood as a cultural construct," *The British Journal of Psychiatry* 184, No. 1 (December 2003): 8–9.

Tinbergen, Niko, *The Study of Instinct* (Oxford: Clarendon Press, 1969).

Tobias, Phillip V., *The Brain in Hominid Evolution* (New York: Columbia University Press, 1971).

Tomasello, Michael, Malinda Carpenter, Joseph Call, Tanya Behne, and Henrike Moll, "Understanding and sharing intentions: the origins of cultural cognition," *Behavioral and Brain Sciences* 28 (2005): 675–735.

Tonnies, Ferdinand, *Gemeinschaft und Gesellschaft* (Darmstadt: Wissenschaftliche Buchgesellschaft, 2005). Originally published in 1887.

Towner, Sarah, "Concept of mind in non-human primates," *Bioscience Horizons* 3, No. 1 (2010): 96–104.

Toynbee, Arnold, *A Study of History* (London: Oxford University Press, 1987).

Tragust, Simon, Heike Feldhaar, Xavier Espadaler, and Jes Soe Pedersen, "Rapid increase of the parasitic fungus *Laboulbenia formicarum* in supercolonies of the invasive garden ant *Lasius neglectus*," *Biological Invasions* 17, No. 10 (October 2015): 2795–2801.

Treffert, Darold A., *Extraordinary People: Understanding Savant Syndrome* (New York: Harper and Row and Ballantine, 1989).

Trevors, J.T., "Evolution of gene transfer in bacteria," *World Journal of Microbiology and Biotechnology* 15, No. 1 (February 1999): 1–6.

Trivers, R.L., "Parental investment and sexual selection," in *Sexual Selection and the Descent of Man,* edited by B. Campbell (Chicago: Aldine, 1972): 136–179.

Trinkaus, Eric, "Neandertals, early modern humans, and rodeo riders," *Journal of Archaeological Science* 39, No. 12 (December 2012): 3691–3693.

Trinkaus, J.P., *Cells into Organs: The Forces That Shape the Embryo* (Englewood Cliffs, NJ: Prentice-Hall, 1969).
Tsang, Y.C., "The functions of the visual areas of the cerebral cortex of the rat in the learning and retention of the maze I," *Comparative Psychology Monographs* 10, No. 4 (1934): 1–56.
Tsang, Y.C., "The functions of the visual areas of the cerebral cortex of the rat in the learning and retention of the maze II," *Comparative Psychology Monographs* 12, No. 2 (1936): 1–41.
Tucci, Debra L., "Nose and sinuses," Merck Manual Consumer Version (2016). http://www.merckmanuals.com/home/ear,-nose,-and-throat-disorders/biology-of-the-ears,-nose,-and-throat/nose-and-sinuses, accessed January 7, 2017.
Tucker, Patrick, *The Naked Future: What Happens in a World That Anticipates Your Every Move?* (New York: Current Books, 2014).
Tucker, William, *Marriage and Civilization: How Monogamy Made Us Human* (Washington, DC: Regnery, 2014).
Turkheimer, E., A. Haley, M. Waldron, B. D'Onofrio, and I.I. Gottesman, "Socioeconomic status modifies heritability of IQ in young children," *Psychological Science* 14, No. 6 (2003): 623–628.
Turner, William, "The convolutions of the brain: a study in comparative anatomy," Address to the Anatomical Section of the 10th International Medical Congress in Berlin, Tuesday, August 6, 1890.
Tzar, Jennifer, and Eric Scigliano, "Through the eye of an octopus," *Discover Magazine*, October 2003.
Ugelvig, Line V., and Sylvia Cremer, "Effects of social immunity and unicoloniality on host–parasite interactions in invasive insect societies," *Functional Ecology* 26, No. 6 (December 2012): 1300–1312.
Ungar, Peter S., ed., *Evolution of the Human Diet: The Known, the Unknown, and the Unknowable* (Oxford: Oxford University Press, 2007).
Uylings, B.M., Henk J. Groenewegen, and Bryan Kolb, "Do rats have a prefrontal cortex?" *Behavioral Brain Research* 146 (2003): 3–17.
Van Dongen, P.A.M., "Brain size in vertebrates," in *The Central Nervous System of Vertebrates*, edited by R. Nieuwenhuys, H.J. ten Donkelaar, and C. Nicholsen (Berlin: Springer, 1998): 2099–2134.
Van Hoof, Anton, *From Autothanasia to Suicide* (London: Routledge, 1990).
Van Schaik, Carel P., Karin Isler, and Judith M. Burkart, "Explaining brain size variation: from social to cultural brain," *Trends in Cognitive Sciences* 16, No. 5 (May 16, 2012): 277–284.
Vanderbilt, Tom, "The traffic guru," *The Wilson Quarterly*, Summer 2008.
Vekua, Abesalom, David Lordkipanidze, G.P. Rightmire et al., "A new skull of early Homo from Dmanisi, Georgia," *Science* 297, No. 5578 (2002): 85–89.
Velando, Alberto, Rene Beamonte-Barrientos, and Roxana Torres, "Pigment-based skin colour in the blue-footed Booby: an honest signal of current condition used by females to adjust reproductive investment," *Oecologia* 149 (2006): 535–542. doi: 10.1007/s00442–006–0457–5.
Venticinque, Eduardo M., Harold G. Fowler, and Carlos A. Silva, "Modes and frequencies of colonization and its relation to extinctions, habitat and seasonality in

the social spider *Anelosimus eximius* in the Amazon (Araneidae: Theridiidae)," *Psyche* 100 (1993): 35–41.

Verrelli, B.C., and S.A. Tishkoff, "Signatures of selection and gene conversion associated with human color vision variation," *American Journal Human Genetics* 75 (2004): 363–375.

Villa, Paola, *Terra Amata and the Middle Pleistocene Record of Southern France* (Berkeley: University of California Press, 1983).

Visscher, P.K., and T.D. Seeley, "Foraging strategy of honeybee colonies in a temperate deciduous forest," *Ecology* 63 (1982): 1790–1801.

Visscher, P. Kirk, "Group decision making in nest-site selection among social insects," *Annual Review of Entomology* 52 (2007): 255–275.

Vlcek, Emanuel, "The fossil man of Ganovce, Czechoslovakia," *Journal of the Royal Anthropological Institute of Great Britain and Ireland* 85, No. ½ (1955): 163–172.

Volkow, Nora D., Gene-Jack Wang, and Scott H. Kollins, "Evaluating dopamine reward pathway in ADHD," *JAMA* 302, No. 10 (September 9, 2009): 1084–1091.

Von Harnack, Adolf, *What Is Christianity?* (London: Williams and Norgate, 1901).

Vyse, Stuart, *Believing in Magic: The Psychology of Superstition* (New York: Oxford University Press, 1997).

Waddington, C.H., "Evolutionary systems—animal and human," *Nature* 183 (June 13, 1959): 1634–1638.

Walker, L., H. Levine, and M. Jucker, "Koch's postulates and infectious proteins," *Acta Neuropathologica* 112, No. 1 (2006): 1–4.

Wallace, Anthony F.C., "Mental illness, biology and culture," in *Psychological Anthropology*, edited by Francis Hsu (Cambridge: Schenkman, 1972): 363–402.

Wallis, J.D., K.C. Anderson, and E.K. Miller, "Single neurons in prefrontal cortex encode abstract rules," *Nature* 411 (June 21, 2001): 953–956.

Wang, Yan, Yuichi Michikawa, Con Mallidis et al., "Muscle-specific mutations accumulate with aging in critical human mtDNA control sites for replication," *Proceedings of the National Academy of Sciences, USA* 98, No. 7 (2001): 4022–4027.

Wasman, E., "Neue Beitrage zur Biologie von Lomechusa und Atemeles, mit kritischen Bemerkungen uber das echte Gastverhalthnis," *Zeitschrift fur Wissenschaftliche Zoologie* 114, No. 2 (1915): 233–402.

Waterhouse, Lynn, "Multiple intelligences, the Mozart effect, and emotional intelligence: a critical review," *Educational Psychologist* 41, No. 4 (2006): 207–225.

Waters, Rob, "Alert on antidepressants for kids panel urges sterner FDA warnings on children's suicide risks," *San Francisco Chronicle*, February 3, 2004.

Weaver, Thomas, and Douglas White, editors, *The Anthropology of Urban Environments* (Washington, DC: Society for Applied Anthropology, 1972).

Webb, Barbara, "Cognition in insects," *Philosophical Transactions of the Royal Society of London B* 367 (2012): 2715–2722.

Webb, Sidney, and Beatrice Webb, *English Poor Law History* (London: Longmans, 1927).

Weber, Max, *The Protestant Ethic and the Spirit of Capitalism*, translated by Talcott Parsons with notes (New York: Scribner, 1948 edition).

Weber, Max, *The City*, translated and edited by Don Martindale and Grutrud Neuwirth (New York: The Free Press, 1958).

Weber, Max, *Confucianism and Taoism* (New York: The Free Press, 1959).

Wedekind, C., T. Seebeck, F. Bettens, A.J. Paepke, "MHC-dependent mate preferences in humans." *Proceedings of the Royal Society of London B: Biological Sciences* 260, No. 1359 (June 22, 1995): 245–249.

Wehner, Rudiger, Tsukasa Fukushi, and Karin Isler, "On bing small: brain allometry in ants," *Brain Behavior and Evolution* 69 (2007): 220–228.

Weidenreich, Franz, "The duration of life of fossil man in China and the pathological legions found in his skeleton," *Chinese Medical Journal* 55 (January 1939): 34–44.

Weidenreich, Franz, *Shorter Anthropological Papers of Franz Weidenreich Published in the Period 1939–1948* (New York: Viking Fund, 1949).

Weiner, Norbert, *Cybernetics* (Paris: Herman et cie, 1948).

Weiner, Norbert, *The Human Use of Human Beings* (Boston: Houghton Mifflin, 1954).

Weismann, Adam, and Katy Brice, *Building with Cob* (London: Green Books, 2006).

Weiss, K.M., "Evolutionary perspectives on human aging," in *Other Ways of Growing Old*, edited by P. Amoss and S. Harrel (Stanford: Stanford University Press, 1981): 25–28.

Weissman, C., "The state of the prion," *Nature Reviews in Microbiology* 2 (2004): 861–871.

Wells, Calvin, *Bones, Bodies and Disease: Evidence of Disease and Abnormality in Early Man* (New York: Frederick A. Praeger, 1964).

Wells, Jonathan, *The Evolutionary Biology of Human Body Fatness* (Cambridge: Cambridge University Press, 2010).

Wendt, Herbert, *In Search of Adam* (New York: Collier Books, 1963).

Wesotowski, Tomasz, "Reduction of phallus in birds, an avian way to safe sex?" *Journal of Avian Biology* 30, No. 4 (December 1999): 483–485.

West, Geoffrey, *Scale: The Universal Laws of Growth, Innovation, Sustainability, and the Pace of Life in Organisms, Cities, Economies, and Companies* (New York: Penguin Press, 2017).

Weston, Eleanor M., and Adrian M. Lister, "Insular dwarfism in hippos and a model for brain size reduction in *Homo floresiensis*," *Nature* 459, No. 7243 (May 7, 2009): 85–88.

Weyer, Edward, *The Eskimos: Their Environment and Folkways* (Hamden, CT: Archon Books, 1932).

White, L., "The locus of mathematical reality: an anthropological footnote," *Philosophy of Science* 14, No. 4 (1947): 289–303.

White, Leslie A., *A Science of Culture: The Study of Man and Civilization* (New York: Grove Press, 1949).

White, Tim, *Prehistoric Cannibalism at Mancos 5MTUMR-2346* (Princeton: Princeton University Press, 1992).

Widd, Thomas, *The Deaf and Dumb and Blind Deaf-Mute, with Interesting Facts and Anecdotes: A Short History of the MacKay Institution* (Montreal: K.K. Grafton, 1880).

Wiens, John A., "An approach to the study of ecological relationships among grassland birds," Dissertation, original from University of Wisconsin–Madison (Ann Arbor, MI: University Microfilms, 1966).

Wilkins, Adam S., and Robin Holliday, "The evolution of meiosis from mitosis," *Genetics* 181 (2009): 3–12.

Williams, G.C., *Sex and Evolution* (Princeton, NJ: Princeton University Press, 1975).

Williams, S.A., and M.R. Shattuck, "Ecology, longevity and naked mole-rats: confounding effects of sociality?" *Proceedings of the Royal Society of London B: Biological Sciences* 282, No. 1802 (2015).

Williams, Trevor I., *The History of Invention from Stone Axes to Silicon Chips* (New York: Facts on File, 1987).

Williams, Vernon J., Jr., *Rethinking Race: Franz Boas and His Contemporaries* (Lexington: University Press of Kentucky, 1996).

Willingham, Daniel T., "Understanding ADHD," *American Educator* (Winter 2004/2005): 36–41.

Wilson, E.O., *The Insect Societies* (Cambridge: Belknap Press of Harvard University Press, 1971).

Wilson, E.O., *Sociobiology: A New Synthesis* (Cambridge: Belknap Press of Harvard University Press, 1975).

Wilson, E.O., *The Social Conquest of the Earth* (New York: Liveright, 2012).

Wilson, E.O., and Bert Hölldobler, "Eusociality: origin and consequences," *Proceedings of the National Academy of Sciences, USA* 102, No. 38 (September 20, 2005): 13367–13371.

Wilson, E.O., and Bert Hölldobler, *The Superorganism: The Beauty, Elegance, and Strangeness of Insect Societies* (New York: W.W. Norton, 2009).

Wilson, E.O., and M.A. Norwak, "Natural selection drives the evolution of ant life cycles," *Proceedings of the National Academy of Sciences, USA* 111 (2014): 12585–12590.

Wilson, D.S., and E. Sober, "Reviving the superorganism." *Journal of Theoretical Biology* 136 (1989): 337–356.

Wilson, John A., *The Culture of Ancient Egypt* (Chicago: University of Chicago Press, 1951).

Wilson, Peter J., *The Domestication of the Human Species* (New Haven: Yale University Press, 1988).

Wilsson, Lars, *My Beaver Colony* (New York: Doubleday, 1968).

Wilsson, Lars, "Observations and experiment on the ethology of the European beaver *(Castor fiber L.)," Viltrevy Swedish Wildlife* 8 (1971): 31–266.

Wireman, J.W., and M. Dworkin, "Morphogenesis and developmental interactions in Myxobacteria," *Science* 189 (1975): 516–523.

Wittgenstein, Ludwig, *Philosophical Investigations,* 3rd edition (New York: Macmillan, 1953).

Wittfogel, Karl, *Oriental Despotism: A Comparative Study of Total Power* (New York: Random House, 1957).

Woodburne, Michael O., Gregg F. Gunnell, and Richard K. Stucky, "Climate directly influences Eocene mammal faunal dynamics in North America," *Proceedings of the National Academy of Sciences, USA* 106, No. 32 (August 11, 2009): 13399–13403.

Woodley, Michael A., Heiner Rinderman, Edward Bell, James Stratford, and Davide Piffer, "The relationship between microcephalin, ASPM and intelligence: a reconsideration," *Intelligence* 44 (May–June 2014): 51–63.

Woods, C.G., "Human microcephaly," *Current Opinion in Neurobiology* 14 (2004): 112–117.

Woolfolk, Robert L., "Malfunction and mental illness," *The Monist* 82, No. 4, Cognitive Theories of Mental Illness (October 1999): 658–670.

Wooton, Raymond J., and David F. Allen, "Dramatic religious conversion and schizophrenic decompensation," *Journal of Religion and Health* 22, No. 3 (1983): 212–220.

Worster, David, *The Ends of the Earth* (Cambridge: Cambridge University Press, 1988).

Wright, Robert, *The Moral Animal: Why We Are, the Way We Are: The New Science of Evolutionary Psychology* (New York: Vintage, 1995).

Wynn, T., "The intelligence of Later Acheulean hominids," *Man*, New Series 14, No. 3 (September 1979): 371–391.

Wynn, T., and W.C. McCrew, "An ape's view of the Oldowan," *Man*, New Series 24, No. 3 (1989): 383–398.

Wynne-Edwards, V.C., "Intergroup selection in the evolution of social systems," *Nature* 200, No. 4907 (November 16, 1963): 623–636.

Wystrach, Antoine, Sebastian Schwarz, Alice Baniel, and Ken Cheng, "Backtracking behaviour in lost ants: an additional strategy in their navigational toolkit," *Proceedings of the Royal Society B: Biological Sciences* (August 21, 2013). doi: 10.1098/rspb.2013.1677.

Wystrach, Antoine, "We've been looking at ant intelligence the wrong way," *Scientific American*, "The Conversation" (August 30, 2013).

Xinhua (New China News Agency), "China's Urban Living Space per Capita Reaches 25 Square Meters," August 5, 2005.

Yang, Charles, "Ontogeny and phylogeny of language," *Proceedings of the National Academy of Sciences, USA* 110, No. 16 (April 16, 2013): 6324–6327.

Yip, Eric C., Kimberly S. Powers, and Leticia Aviles, "Cooperative capture of large prey solves scaling challenge faced by spider societies," *Proceedings of the National Academy of Sciences, USA* 105, No. 33 (August 19, 2009): 11818–11822.

Young, J.Z., *The Anatomy of the Nervous System of Octopus vulgaris* (Oxford: Clarendon Press, 1971).

Yuan, Gang, and Bing Zhu, "Histone variants and epigenetic inheritance," *Biochimica et Biophysica Acta (BBA) – Gene Regulatory Mechanisms* 1819, No. 3–4 (March–April 2012): 222–229.

Zangwill, O.L., "Lashley's concept of cerebral mass action," in *Current Problems in Animal Behavior*, edited by W.H. Thorpe and O.L. Zangwill (Cambridge: Cambridge University Press, 1961): 59–86.

Zeiderman, Austin, "Risk, rights and resettlement," *Anthropology News* (December 2008: 9–10).

Zhang, Shaowu, Fiola Bock, Aung Si et al., "Visual working memory in decision making by honey bees," *Proceedings of the National Academy of Sciences, USA* 102, No. 14 (April 5, 2005): 5250–5255.

Zhukovskaya, Marianna, Aya Yanagawa, and Brian T. Forschler, "Grooming behavior as a mechanism of insect disease defense," *Insects* 4 (2013): 609–630. doi: 10.3390/insects4040609.

Zilles, K., N. Palomero-Gallagher, and K. Amunts, "Development of cortical folding during evolution and ontogeny," *Trends in Neurosciences* 36 (2013): 275–284.

Zuckerman, B., J. Stevenson, and V. Bailey, "Sleep problems in early childhood: continuities, predictive factors and behavioral correlates," *Pediatrics* 30 (1987): 664–671.

Zuk, Marlene, *Paleofantasy: What Evolution Really Tells Us About Sex, Diet and How We Live* (New York: W.W. Norton, 1987).

Zullo, Letizia, German Sumbre, Claudio Agnisola, Tamar Flash, and Binyamin Hochner, "Nonsomatotopic organization of the higher motor centers in octopus," *Current Biology* 19, No. 19 (October 13, 2009): 1632–1636.

Zurowski, W., "Building activity of beavers," *Acta Theriologica* 37, No. 4 (1992): 403–411.

Index

About the Author

Niccolo Leo Caldararo, PhD, is a lecturer who teaches in the anthropology department of San Francisco State University. His area of research is human evolution, paleoanthropology, economic anthropology, religion, and psychological anthropology. He has made more than 20 ethnographic studies for public and private institutions and firms and published widely on the evolution of hominids and economic anthropology, as well as the anthropology of religion. Introduced by Carlo Cipolla to the evolution of financial institutions while a student at the University of California at Berkeley in the 1960s, he pursued comparative economic studies of traditional peoples in various parts of the world including Central America, Polynesia, and Melanesia. Hired as a laboratory assistant by J. Desmond Clark in 1967, he began a career studying and preparing fossils and artifacts for study and the biological adaptations of primates and human variability.

www.ingramcontent.com/pod-product-compliance
Ingram Content Group UK Ltd.
Pitfield, Milton Keynes, MK11 3LW, UK
UKHW041126071026
15587UKWH00009B/243

9 781498 540872